FOUNDATION
COMPUTAT

FOUNDATIONS OF COMPUTATION

BY

ELLA HALL
RUBY JAMES

KNOWLEDGE
BAKERS

KNOWLEDGE BAKERS

ISBN: 978-93-9001-362-3 (HB)

Published by : **Knowledge Bakers**
Pune, India
Email: info@knowledgebakers.com

PREFACE

Computation serves as the bedrock of modern technological advancement, underpinning innovations in fields as diverse as artificial intelligence, data science, cryptography, and software engineering. At its core, computation involves the manipulation of abstract symbols according to predefined rules, enabling the solution of complex problems, the execution of algorithms, and the creation of digital systems that permeate every aspect of our lives. As we embark on this journey through the foundations of computation, it becomes evident that understanding the fundamental principles and concepts is essential for unlocking the full potential of computational thinking and harnessing its transformative power in various domains.

This book, titled "Foundations of Computation," is designed to provide a comprehensive introduction to the theoretical underpinnings, mathematical models, and practical applications of computation. Through collaboration with leading experts and educators in the field, we aim to offer readers a rigorous yet accessible exploration of key topics, from formal languages and automata theory to computational complexity, algorithms, and beyond.

As editors of this volume, we have curated a collection of chapters that cover a wide range of foundational concepts and techniques in computation, presenting each topic in a clear and structured manner. Each chapter is structured to provide a conceptual framework for understanding key principles, followed by illustrative examples, problem-solving exercises, and real-world applications to reinforce learning and facilitate deeper comprehension.

Furthermore, this book seeks to highlight the interdisciplinary nature of computation and its connections to mathematics, logic, computer science, and beyond. By showcasing examples of how computational thinking is applied in diverse fields, from cryptography and bioinformatics to computer graphics and natural language processing, we aim to inspire readers to develop a holistic understanding of computation and its potential for driving innovation and addressing complex challenges in the digital age.

As we embark on this enlightening journey through the foundations of computation, we extend our gratitude to the contributors whose expertise and dedication have enriched this volume. We hope that "Foundations of Computation" serves as a valuable resource for students, educators, and practitioners alike, empowering them to develop a solid foundation in computational thinking and to explore new frontiers in the ever-evolving landscape of computation and technology.

Ella Hall
Ruby James

Contents

Chapter 1

Logic and Proof

In a sense, we know a lot more than we realize, because everything that we know has consequences—*logical* consequences—that follow automatically. If you know that all humans are mortal, and you know that you are human, then in a sense you know that you are mortal, whether or not you have ever considered or wanted to consider that fact. This is an example of ***logical deduction***: From the ***premises*** that "All humans are mortal" and "I am human," the ***conclusion*** that "I am mortal" can be deduced by logic.

Logical deduction is a kind of computation. By applying rules of logic to a given set of premises, conclusions that follow from those premises can be generated automatically, by a computational process which could be carried out by a computer. Once you know the premises, or are willing to accept them for the sake of argument, you are forced—by logic—to accept the conclusions. Still, to say that you "know" those conclusions would be misleading. The problem is that there are too many of them (infinitely many), and, in general, most of them are not particularly interesting. Until you have actually made the deduction, you don't *really* know the conclusion, and knowing which of the possible chains of deduction to follow is not easy. The *art* of logic is to find an interesting conclusion and a chain of logical deductions that leads from the premises to that conclusion. Checking that the deductions are valid is the mechanical, computational side of logic.

This chapter is mostly about the mechanics of logic. We will investigate logic as a branch of mathematics, with its own symbols, formulas, and rules of computation. Your object is to learn the rules of logic, to understand why they are valid, and to develop skill in applying them. As with any branch of mathematics, there is a certain beauty to the symbols and formulas themselves. But it is the applications that bring the subject to life for most people. We will, of course, cover some applications as we go along. In a sense, though, the real applications of logic include much of computer science and of mathematics itself.

Among the fundamental elements of thought, and therefore of logic, are propositions. A ***proposition*** is a statement that has a truth value: It is either true or false. "Grass is green" and "$2 + 2 = 5$" are propositions. In the first part of this chapter, we will study ***propositional logic***, which takes propositions as basic and considers how they can be combined and manipulated. This branch of logic has surprising application to the design of the electronic circuits that make up computers.

Logic gets more interesting when we consider the internal structure of propositions. In English, a proposition is expressed as a sentence, and, as you know from studying grammar, sentences have parts. A simple sentence like "Grass is green" has a ***subject*** and a ***predicate***. The sentence says something about its subject. The subject of "Grass is green" is grass. The sentence says something about grass. The *something* that the sentence says about its subject is the predicate. In the example, the predicate is the phrase "is green." Once we start working with predicates, we can create propositions using ***quantifiers*** like "all," "some," and "no." For example, working with the predicate "is above average," we can move from simple propositions like "Johnny is above average" to "All children are above average" or to "No child is above average" or to the rather more realistic "Some children are above average." Logical deduction usually deals with quantified statements, as shown by the basic example of human mortality with which we began this chapter. Logical deduction will be a major topic of this chapter; under the name of ***proof***, it will be the last major topic of this chapter, and a major tool for the rest of this book.

1.1 Propositional Logic

A proposition is a statement which is either true or false. In propositional logic, we take propositions as basic and see what we can do with them. Since this is mathematics, we need to be able to talk about propositions without saying which particular propositions we are talking about, so we use symbolic names to represent them. We will always use lowercase letters such as p, q, and r to represent propositions. A letter used in this way is called a ***propositional variable***. Remember that when I say something like "Let p be a proposition," I mean "For the rest of this discussion, let the symbol p stand for some particular statement, which is either true or false (although I am not at the moment making any assumption about which it is)." The discussion has ***mathematical generality*** in that p can represent any statement, and the discussion will be valid no matter which statement it represents.

What we do with propositions is combine them with ***logical operators***. A logical operator can be applied to one or more propositions to produce a new proposition. The truth value of the new proposition is completely determined by the operator and by the truth values of the propositions to which it is applied.[1] In English, logical operators are represented by words such as "and," "or," and "not." For example, the proposition "I wanted to leave and I left" is formed from two simpler propositions joined by the word "and." Adding the word "not" to the proposition "I left" gives "I did not leave" (after a bit of necessary grammatical adjustment).

But English is a little too rich for mathematical logic. When you read the sentence "I wanted to leave and I left," you probably see a connotation of causality: I left *because* I wanted to leave. This implication does not follow from the logical combination of the truth values of the two propositions "I wanted to leave" and "I left." Or consider the proposition "I wanted to leave but I did not leave." Here, the word "but" has the same *logical* meaning as the word "and," but the connotation is very different. So, in mathematical logic, we use *symbols* to represent logical operators. These symbols do not carry any connotation beyond their defined logical meaning. The logical operators corresponding to the English words "and," "or," and "not" are $\wedge$, $\vee$, and $\neg$.

Definition 1.1. Let p and q be propositions. Then $p \vee q$, $p \wedge q$, and $\neg p$ are propositions, whose truth values are given by the rules:

- $p \wedge q$ is true when both p is true and q is true, and in no other case.
- $p \vee q$ is true when either p is true, or q is true, or both p and q are true, and in no other case.
- $\neg p$ is true when p is false, and in no other case.

The operators $\wedge$, $\vee$, and $\neg$ are referred to as ***conjunction***, ***disjunction***, and ***negation***, respectively. (Note that $p \wedge q$ is read as "p and q," $p \vee q$ is read as "p or q," and $\neg p$ is read as "not p.")

[1]It is not always true that the truth value of a sentence can be determined from the truth values of its component parts. For example, if p is a proposition, then "Sarah Palin believes p" is also a proposition, so "Sarah Palin believes" is some kind of operator. However, it does not count as a *logical* operator because just from knowing whether or not p is true, we get no information at all about whether "Sarah Palin believes p" is true.

These operators can be used in more complicated expressions, such as $p \wedge (\neg q)$ or $(p \vee q) \wedge (q \vee r)$. A proposition made up of simpler propositions and logical operators is called a ***compound proposition***. Parentheses can be used in compound expressions to indicate the order in which the operators are to be evaluated. In the absence of parentheses, the order of evaluation is determined by ***precedence rules***. For the logical operators defined above, the rules are that $\neg$ has higher precedence that $\wedge$, and $\wedge$ has precedence over $\vee$. This means that in the absence of parentheses, any $\neg$ operators are evaluated first, followed by any $\wedge$ operators, followed by any $\vee$ operators.

For example, the expression $\neg p \vee q \wedge r$ is equivalent to the expression $(\neg p) \vee (q \wedge r)$, while $p \vee q \wedge q \vee r$ is equivalent to $p \vee (q \wedge q) \vee r$. As a practical matter, when you make up your own expressions, it is usually better to put in parentheses to make your meaning clear. Remember that even if you leave out parentheses, your expression has an unambiguous meaning. If you say "$\neg p \wedge q$" when what you meant was "$\neg(p \wedge q)$," you've got it wrong!

This still leaves open the question of which of the $\wedge$ operators in the expression $p \wedge q \wedge r$ is evaluated first. This is settled by the following rule: When several operators of equal precedence occur in the absence of parentheses, they are evaluated from left to right. Thus, the expression $p \wedge q \wedge r$ is equivalent to $(p \wedge q) \wedge r$ rather than to $p \wedge (q \wedge r)$. In this particular case, as a matter of fact, it doesn't really matter which $\wedge$ operator is evaluated first, since the two compound propositions $(p \wedge q) \wedge r$ and $p \wedge (q \wedge r)$ always have the same value, no matter what logical values the component propositions p, q, and r have. We say that $\wedge$ is an ***associative*** operation. We'll see more about associativity and other properties of operations in the next section.

Suppose we want to verify that, in fact, $(p \wedge q) \wedge r$ and $p \wedge (q \wedge r)$ do always have the same value. To do so, we have to consider all possible combinations of values of p, q, and r, and check that for all such combinations, the two compound expressions do indeed have the same value. It is convenient to organize this computation into a truth table. A ***truth table*** is a table that shows the value of one or more compound propositions for each possible combination of values of the propositional variables that they contain. Figure 1.1 is a truth table that compares the value of $(p \wedge q) \wedge r$ to the value of $p \wedge (q \wedge r)$ for all possible values of p, q, and r. There are eight rows in the table because there are exactly eight different ways in which truth values can be assigned to p, q, and r.[2] In this table, we see that the

[2]In general, if there are n variables, then there are 2^n different ways to assign truth values to the variables. This might become clear to you if you try to come up with a

p	q	r	$p \wedge q$	$q \wedge r$	$(p \wedge q) \wedge r$	$p \wedge (q \wedge r)$
false	false	false	false	false	false	false
false	false	true	false	false	false	false
false	true	false	false	false	false	false
false	true	true	false	true	false	false
true	false	false	false	false	false	false
true	false	true	false	false	false	false
true	true	false	true	false	false	false
true	true	true	true	true	true	true

Figure 1.1: A truth table that demonstrates the logical equivalence of $(p \wedge q) \wedge r$ and $p \wedge (q \wedge r)$. The fact that the last two columns of this table are identical shows that these two expressions have the same value for all eight possible combinations of values of p, q, and r.

last two columns, representing the values of $(p \wedge q) \wedge r$ and $p \wedge (q \wedge r)$, are identical.

More generally, we say that two compound propositions are ***logically equivalent*** if they always have the same value, no matter what truth values are assigned to the propositional variables that they contain. If the number of propositional variables is small, it is easy to use a truth table to check whether or not two propositions are logically equivalent.

There are other logical operators besides $\wedge$, $\vee$, and $\neg$. We will consider the ***conditional operator***, $\rightarrow$, the ***biconditional operator***, $\leftrightarrow$, and the ***exclusive or operator***, $\oplus$.[3] These operators can be completely defined by a truth table that shows their values for the four possible combinations of truth values of p and q.

Definition 1.2. For any propositions p and q, we define the propositions $p \rightarrow q$, $p \leftrightarrow q$, and $p \oplus q$ according to the truth table:

---

scheme for systematically listing all possible sets of values. If not, you'll find a rigorous proof of the fact later in this chapter.

[3]Note that the symbols used in this book for the logical operators are not universal. While $\wedge$, $\vee$, and $\rightarrow$ are fairly standard, $\neg$ is often replaced by $\sim$ and $\leftrightarrow$ is sometimes represented by $\equiv$ or $\Leftrightarrow$. There is even less standardization of the exclusive or operator, but that operator is generally not so important as the others.

p	q	$p \to q$	$p \leftrightarrow q$	$p \oplus q$
false	false	true	true	false
false	true	true	false	true
true	false	false	false	true
true	true	true	true	false

When these operators are used in expressions, in the absence of parentheses to indicate order of evaluation, we use the following precedence rules: The exclusive or operator, $\oplus$, has the same precedence as $\vee$. The conditional operator, $\to$, has lower precedence than $\wedge$, $\vee$, $\neg$, and $\oplus$, and is therefore evaluated after them. Finally, the biconditional operator, $\leftrightarrow$, has the lowest precedence and is therefore evaluated last. For example, the expression "$p \to q \wedge r \leftrightarrow \neg p \oplus s$" is evaluated as if it were written "$(p \to (q \wedge r)) \leftrightarrow ((\neg p) \oplus s)$."

In order to work effectively with the logical operators, you need to know more about their meaning and how they relate to ordinary English expressions.

The proposition $p \to q$ is called an ***implication*** or a ***conditional***. It is usually read as "p implies q." In English, $p \to q$ is often expressed as "if p then q." For example, if p represents the proposition "Bill Gates is poor" and q represents "the moon is made of green cheese," then $p \to q$ could be expressed in English as "If Bill Gates is poor, then the moon is made of green cheese." In this example, p is false and q is also false. Checking the definition of $p \to q$, we see that $p \to q$ is a true statement. Most people would agree with this. It's worth looking at a similar example in more detail. Suppose that I assert that "If the Mets are a great team, then I'm the king of France." This statement has the form $m \to k$ where m is the proposition "the Mets are a great team" and k is the proposition "I'm the king of France." Now, demonstrably I am not the king of France, so k is false. Since k is false, the only way for $m \to k$ to be true is for m to be false as well. (Check the definition of $\to$ in the table!) So, by asserting $m \to k$, I am really asserting that the Mets are *not* a great team.

Or consider the statement, "If the party is on Tuesday, then I'll be there." What am I trying to say if I assert this statement? I am asserting that $p \to q$ is true, where p represents "The party is on Tuesday" and q represents "I will be at the party." Suppose that p is true, that is, the party does in fact take place on Tuesday. Checking the definition of $\to$, we see that in the only case where p is true and $p \to q$ is true, q is also true. So from the truth of "If the party is on Tuesday, then I will be at the party" and "The party is in fact on Tuesday," you can deduce that "I will be at the party" is also true. But suppose, on the other hand, that the party is actually on Wednesday. Then p is false. When p is false and $p \to q$ is true, the definition of $p \to q$ allows q to be either true or false. So, in this case, you can't make any deduction about whether or not I will be at the party. The statement "If the party is on Tuesday, then I'll be there" doesn't assert anything about what will happen if the party is on some other day than Tuesday.

The implication $(\neg q) \to (\neg p)$ is called the ***contrapositive*** of $p \to q$. An implication is logically equivalent to its contrapositive. The contrapositive of "If this is Tuesday, then we are in Belgium" is "If we aren't in Belgium, then this isn't Tuesday." These two sentences assert exactly the same thing.

Note that $p \to q$ is *not* logically equivalent to $q \to p$. The implication $q \to p$ is called the ***converse*** of $p \to q$. The converse of "If this is Tuesday, then we are in Belgium" is "If we are in Belgium, then this is Tuesday." Note that it is possible for either one of these statements to be true while the other is false. In English, I might express the fact that both statements are true by saying "If this is Tuesday, then we are in Belgium, *and conversely.*" In logic, this would be expressed with a proposition of the form $(p \to q) \land (q \to p)$.

The biconditional operator is closely related to the conditional operator. In fact, $p \leftrightarrow q$ is logically equivalent to $(p \to q) \land (q \to p)$. The proposition $p \leftrightarrow q$ is usually read as "p if and only if q." (The "p if q" part represents $q \to p$, while "p only if q" is another way of asserting that $p \to q$.) It could also be expressed as "if p then q, and conversely." Occasionally in English, "if... then" is used when what is really meant is "if and only if." For example, if a parent tells a child, "If you are good, Santa will bring you toys," the parent probably really means to say "Santa will bring you toys if and only if you are good." (The parent would probably not respond well to the child's perfectly logical plea "But you never said what would happen if I wasn't good!")

Finally, we turn to the exclusive or operator. The English word "or" is actually somewhat ambiguous. The two operators $\oplus$ and $\lor$ express the two possible meanings of this word. The proposition $p \lor q$ can be expressed unambiguously as "p or q, *or both,*" while $p \oplus q$ stands for "p or q, *but not both.*" If a menu says that you can choose soup or salad, it doesn't mean that you can have both. In this case, "or" is an exclusive or. On the other hand, in "You are at risk of heart disease if you smoke or drink," the or is inclusive since you certainly don't get off the hook if you both smoke and drink. In mathematics, the word "or" is always taken in the inclusive sense of $p \lor q$.

Now, any compound proposition that uses any of the operators $\to$, $\leftrightarrow$, and $\oplus$ can be rewritten as a logically equivalent proposition that uses only $\land$, $\lor$, and $\neg$. It is easy to check that $p \to q$ is logically equivalent to $(\neg p) \lor q$.

(Just make a truth table for $(\neg p) \vee q$.) Similarly, $p \leftrightarrow q$ can be expressed as $((\neg p) \vee q) \wedge ((\neg q) \vee p)$, So, in a strict logical sense, $\rightarrow$, $\leftrightarrow$, and $\oplus$ are unnecessary. (Nevertheless, they are useful and important, and we won't give them up.)

Even more is true: In a strict logical sense, we could do without the conjunction operator $\wedge$. It's easy to check that $p \wedge q$ is logically equivalent to $\neg(\neg p \vee \neg q)$, so any expression that uses $\wedge$ can be rewritten as one that uses only $\neg$ and $\vee$. Alternatively, we could do without $\vee$ and write everything in terms of $\neg$ and $\wedge$.

Certain types of proposition will play a special role in our further work with logic. In particular, we define tautologies and contradictions as follows:

Definition 1.3. A compound proposition is said to be a ***tautology*** if and only if it is *true* for all possible combinations of truth values of the propositional variables which it contains. A compound proposition is said to be a ***contradiction*** if and only if it is *false* for all possible combinations of truth values of the propositional variables which it contains.

For example, the proposition $((p \vee q) \wedge \neg q) \rightarrow p$ is a tautology. This can be checked with a truth table:

p	q	$p \vee q$	$\neg q$	$(p \vee q) \wedge \neg q$	$((p \vee q) \wedge \neg q) \rightarrow p$
false	false	false	true	false	true
false	true	true	false	false	true
true	false	true	true	true	true
true	true	true	false	false	true

The fact that all entries in the last column are true tells us that this expression is a tautology. Note that for any compound proposition P, P is a tautology if and only if $\neg P$ is a contradiction. (Here and in the future, I use uppercase letters to represent compound propositions. P stands for any formula made up of simple propositions, propositional variables, and logical operators.) Logical equivalence can be defined in terms of tautology:

Definition 1.4. Two compound propositions, P and Q, are said to be ***logically equivalent*** if and only if the proposition $P \leftrightarrow Q$ is a tautology.

The assertion that P is logically equivalent to Q will be expressed symbolically as "$P \equiv Q$." For example, $(p \rightarrow q) \equiv (\neg p \vee q)$, and $p \oplus q \equiv (p \vee q) \wedge \neg(p \wedge q)$.

Exercises

1. Give the three truth tables that define the logical operators $\wedge$, $\vee$, and $\neg$.

2. Insert parentheses into the following compound propositions to show the order in which the operators are evaluated:

a) $\neg p \vee q$ **b)** $p \wedge q \vee \neg p$ **c)** $p \vee q \wedge r$ **d)** $p \wedge \neg q \vee r$

3. List the 16 possible combinations of truth values for the four propositional variables s, p, q, r. Try to find a systematic way to list the values. (Hint: Start with the eight combinations of values for p, q, and r, as given in the truth table in Figure 1.1. Now, explain why there are 32 possible combinations of values for five variables, and describe how they could be listed systematically.)

4. Some of the following compound propositions are tautologies, some are contradictions, and some are neither. In each case, use a truth table to decide to which of these categories the proposition belongs:

a) $(p \wedge (p \rightarrow q)) \rightarrow q$ **b)** $((p \rightarrow q) \wedge (q \rightarrow r)) \rightarrow (p \rightarrow r)$
c) $p \wedge (\neg p)$ **d)** $(p \vee q) \rightarrow (p \wedge q)$
e) $p \vee (\neg p)$ **f)** $(p \wedge q) \rightarrow (p \vee q)$

5. Use truth tables to show that each of the following propositions is logically equivalent to $p \leftrightarrow q$.

a) $(p \rightarrow q) \wedge (q \rightarrow p)$ **b)** $(\neg p) \leftrightarrow (\neg q)$
c) $(p \rightarrow q) \wedge ((\neg p) \rightarrow (\neg q))$ **d)** $\neg(p \oplus q)$

6. Is $\rightarrow$ an associative operation? This is, is $(p \rightarrow q) \rightarrow r$ logically equivalent to $p \rightarrow (q \rightarrow r)$? Is $\leftrightarrow$ associative?

7. Let p represent the proposition "You leave" and let q represent the proposition "I leave." Express the following sentences as compound propositions using p and q, and show that they are logically equivalent:

a) Either you leave or I do. (Or both!)
b) If you don't leave, I will.

8. Suppose that m represents the proposition "The Earth moves," c represents "The Earth is the center of the universe," and g represents "Galileo was railroaded." Translate each of the following compound propositions into English:

a) $\neg g \wedge c$ **b)** $m \rightarrow \neg c$
c) $m \leftrightarrow \neg c$ **d)** $(m \rightarrow g) \wedge (c \rightarrow \neg g)$

9. Give the converse and the contrapositive of each of the following English sentences:

a) If you are good, Santa brings you toys.
b) If the package weighs more than one ounce, then you need extra postage.
c) If I have a choice, I don't eat eggplant.

10. In an ordinary deck of fifty-two playing cards, for how many cards is it true

a) that "This card is a ten and this card is a heart"?
b) that "This card is a ten or this card is a heart"?
c) that "If this card is a ten, then this card is a heart"?

d) that "This card is a ten if and only if this card is a heart"?

11. Define a logical operator $\downarrow$ so that $p \downarrow q$ is logically equivalent to $\neg(p \vee q)$. (This operator is usually referred to as "NOR," short for "not or"). Show that each of the propositions $\neg p$, $p \wedge q$, $p \vee q$, $p \rightarrow q$, $p \leftrightarrow q$, and $p \oplus q$ can be rewritten as a logically equivalent proposition that uses $\downarrow$ as its only operator.

1.2 Boolean Algebra

So far we have discussed how to write and interpret propositions. This section deals with *manipulating* them. For this, we need algebra. Ordinary algebra, of the sort taught in high school, is about manipulating numbers, variables that represent numbers, and operators such as + and × that apply to numbers. Now, we need an algebra that applies to logical values, propositional variables, and logical operators. The first person to think of logic in terms of algebra was the mathematician, George Boole, who introduced the idea in a book that he published in 1854. The algebra of logic is now called ***Boolean algebra*** in his honor.

The algebra of numbers includes a large number of rules for manipulating expressions. The distributive law, for example, says that $x(y+z) = xy + xz$, where x, y, and z are variables that stand for any numbers or numerical expressions. This law means that whenever you see something of the form $xy + xz$ in a numerical expression, you can substitute $x(y+z)$ without changing the value of the expression, and *vice versa.* Note that the equals sign in $x(y+z) = xy+xz$ means "has the same value as, no matter what numerical values x, y, and z have."

In Boolean algebra, we work with logical values instead of numerical values. There are only two logical values, true and false. We will write these values as $\mathbb{T}$ and $\mathbb{F}$. The symbols $\mathbb{T}$ and $\mathbb{F}$ play a similar role in Boolean algebra to the role that constant numbers such as 1 and 3.14159 play in ordinary algebra. Instead of the equals sign, Boolean algebra uses logical equivalence, $\equiv$, which has essentially the same meaning.[4] For example, for propositions p, q, and r, the $\equiv$ operator in $p \wedge (q \wedge r) \equiv (p \wedge q) \wedge r$ means "has the same value as, no matter what logical values p, q, and r have."

Many of the rules of Boolean algebra are fairly obvious, if you think a bit about what they mean. Even those that are not obvious can be verified easily by using a truth table. Figure 1.2 lists the most important of these

[4]In ordinary algebra, it's easy to be confused by the equals sign, because it has two very different roles. In an identity such as the distributive law, it means "is always equal to." On the other hand, an equation such as $x^2 + 3x = 4$ is a statement that might or might not be true, depending on the value of x. Boolean algebra has two operators, $\equiv$ and $\leftrightarrow$, that play roles similar to the two roles of the equals sign.

Double negation	$\neg(\neg p) \equiv p$
Excluded middle Contradiction	$p \vee \neg p \equiv \mathbb{T}$ $p \wedge \neg p \equiv \mathbb{F}$
Identity laws	$\mathbb{T} \wedge p \equiv p$ $\mathbb{F} \vee p \equiv p$
Idempotent laws	$p \wedge p \equiv p$ $p \vee p \equiv p$
Commutative laws	$p \wedge q \equiv q \wedge p$ $p \vee q \equiv q \vee p$
Associative laws	$(p \wedge q) \wedge r \equiv p \wedge (q \wedge r)$ $(p \vee q) \vee r \equiv p \vee (q \vee r)$
Distributive laws	$p \wedge (q \vee r) \equiv (p \wedge q) \vee (p \wedge r)$ $p \vee (q \wedge r) \equiv (p \vee q) \wedge (p \vee r)$
DeMorgan's laws	$\neg(p \wedge q) \equiv (\neg p) \vee (\neg q)$ $\neg(p \vee q) \equiv (\neg p) \wedge (\neg q)$

Figure 1.2: Laws of Boolean Algebra. These laws hold for any propositions p, q, and r.

laws. You will notice that all these laws, except the first, come in pairs: Each law in the pair can be obtained from the other by interchanging $\wedge$ with $\vee$ and $\mathbb{T}$ with $\mathbb{F}$. This cuts down on the number of facts you have to remember.[5]

Just as an example, let's verify the first rule in the table, the Law of Double Negation. This law is just the old, basic grammar rule that two negatives make a positive. Although this rule is questionable as it applies to English as it is actually used—no matter what the grammarian says, "I can't get no satisfaction" doesn't really mean "I can get satisfaction"—the validity of the rule in logic can be verified just by computing the two possible cases: when p is true and when p is false. When p is true, then by the definition of the $\neg$ operator, $\neg p$ is false. But then, again by the

[5] It is also an example of a more general fact known as ***duality***, which asserts that given any tautology that uses only the operators $\wedge$, $\vee$, and $\neg$, another tautology can be obtained from it by interchanging $\wedge$ with $\vee$ and $\mathbb{T}$ with $\mathbb{F}$. We won't attempt to prove this here.

definition of $\neg$, the value of $\neg(\neg p)$ is true, which is the same as the value of p. Similarly, in the case where p is false, $\neg(\neg p)$ is also false. Organized into a truth table, this argument takes the rather simple form

p	$\neg p$	$\neg(\neg p)$
true	false	true
false	true	false

The fact that the first and last columns are identical shows the logical equivalence of p and $\neg(\neg p)$. The point here is not just that $\neg(\neg p) \equiv p$, but also that this logical equivalence is valid because it can be verified computationally based just on the relevant definitions. Its validity does *not* follow from the fact that "it's obvious" or "it's a well-known rule of grammar." Students often ask "Why do I have to prove something when it's obvious." The point is that logic—and mathematics more generally—is its own little world with its own set of rules. Although this world is related somehow to the real world, when you say that something is obvious (in the real world), you aren't playing by the rules of the world of logic. The real *magic* of mathematics is that by playing by its rules, you can come up with things that are decidedly not obvious, but that still say something about the real world—often, something interesting and important.

Each of the rules in Figure 1.2 can be verified in the same way, by making a truth table to check all the possible cases.

It's important to understand that the propositional variables in the laws of Boolean algebra can stand for any propositions, including compound propositions. It is not just true, as the Double Negation Law states, that $\neg(\neg p) \equiv p$. It is also true that $\neg(\neg q) \equiv q$, that $\neg(\neg(p \wedge q)) \equiv (p \wedge q)$, that $\neg(\neg(p \rightarrow (q \wedge \neg p))) \equiv (p \rightarrow (q \wedge \neg p))$, and an infinite number of other statements of the same form. Here, a "statement of the same form" is one that can be obtained by substituting something for p in both places where it occurs in $\neg(\neg p) \equiv p$. How can I be sure that all these infinitely many statements are valid when all that I've checked is one little two-line truth table? The answer is that any given proposition, Q, no matter how complicated, has a particular truth value, either true or false. So, the question of the validity of $\neg(\neg Q) \equiv Q$ always reduces to one of the two cases I already checked in the truth table. (Note that for this argument to be valid, the same Q must be substituted for p in every position where it occurs.) While this argument may be "obvious," it is not exactly a proof, but for now we will just accept the validity of the following theorem:

Theorem 1.1 (First Substitution Law). *Suppose that Q is any proposition, and that p is a propositional variable. Consider any tautology. If (Q) is*

substituted for p in all places where p occurs in the tautology, then the result is also a tautology.

Since logical equivalence is defined in terms of tautology, it is also true that when (Q) is substituted for p in a logical equivalence, the result is again a logical equivalence.[6]

The First Substitution Law lets you do algebra! For example, you can substitute $p \to q$ for p in the law of double negation, $\neg(\neg p) \equiv p$. This allows you to "simplify" the expression $\neg(\neg(p \to q))$ to $p \to q$ with confidence that the resulting expression has the same logical value as the expression you started with. (That's what it means for $\neg(\neg(p \to q))$ and $p \to q$ to be logically equivalent.) You can play similar tricks with all the laws in Figure 1.2. Even more important is the Second Substitution Law, which says that you can substitute an expression for a logically equivalent expression, wherever it occurs. Once again, we will accept this as a theorem without trying to prove it here. It is surprisingly hard to put this law into words:

Theorem 1.2 (Second Substitution Law). *Suppose that P and Q are any propositions such that $P \equiv Q$. Suppose that R is any compound proposition in which (P) occurs as a subproposition. Let R' be the proposition that is obtained by substituting (Q) for that occurrence of (P) in R. Then $R \equiv R'$.*

Note that in this case, the theorem does not require (Q) to be substituted for *every* occurrence of (P) in R. You are free to substitute for one, two, or as many occurrences of (P) as you like, and the result is still logically equivalent to R.

The Second Substitution Law allows us to use the logical equivalence $\neg(\neg p) \equiv p$ to "simplify" the expression $q \to (\neg(\neg p))$ by substituting (p) for $(\neg(\neg p))$. The resulting expression, $q \to (p)$, or just $q \to p$ without the parentheses, is logically equivalent to the original $q \to (\neg(\neg p))$. Once again, we have to be careful about parentheses: The fact that $p \vee p \equiv p$ does *not* allow us to rewrite $q \wedge p \vee p \wedge r$ as $q \wedge p \wedge r$. The problem is that $q \wedge p \vee p \wedge r$ means $(q \wedge p) \vee (p \wedge r)$, so that $(p \vee p)$ is not a subexpression. So even though in practice we won't always write all the parentheses, you always have to be aware of where the parentheses belong.

The final piece of algebra in Boolean algebra is the observation that we can chain logical equivalences together. That is, from $P \equiv Q$ and $Q \equiv R$, it follows that $P \equiv R$. This is really just a consequence of the Second

[6] I've added parentheses around Q here for technical reasons. Sometimes, the parentheses are necessary to make sure that Q is evaluated as a whole, so that its final value is used in place of p. As an example of what can go wrong, consider $q \wedge r$. If this is substituted literally for p in $\neg(\neg p)$, without parentheses, the result is $\neg(\neg q \wedge r)$. But this expression means $\neg((\neg q) \wedge r)$, which is *not* equivalent to $q \wedge r$.

Substitution Law: The equivalence $Q \equiv R$ allows us to substitute R for Q in the statement $P \equiv Q$, giving $P \equiv R$. (Remember that, by Definition 1.4, logical equivalence is defined in terms of a proposition.) This means that we can show that two compound propositions are logically equivalent by finding a chain of logical equivalences that lead from one to the other. For example:

$$\begin{aligned} p \wedge (p \rightarrow q) &\equiv p \wedge (\neg p \vee q) && \text{definition of } p \rightarrow q\text{, Theorem 1.2} \\ &\equiv (p \wedge \neg p) \vee (p \wedge q) && \text{Distributive Law} \\ &\equiv \mathbb{F} \vee (p \wedge q) && \text{Law of Contradiction, Theorem 1.2} \\ &\equiv (p \wedge q) && \text{Identity Law} \end{aligned}$$

Each step in the chain has its own justification. In several cases, a substitution law is used without stating as much. In the first line, for example, the definition of $p \rightarrow q$ is that $p \rightarrow q \equiv \neg p \vee q$. The Second Substitution Law allows us to substitute $(\neg p \vee q)$ for $(p \rightarrow q)$. In the last line, we implicitly applied the First Substitution Law to the Identity Law, $\mathbb{F} \vee p \equiv p$, to obtain $\mathbb{F} \vee (p \wedge q) \equiv (p \wedge q)$.

The chain of equivalences in the above example allows us to conclude that $p \wedge (p \rightarrow q)$ is logically equivalent to $p \wedge q$. This means that if you were to make a truth table for these two expressions, the truth values in the column for $p \wedge (p \rightarrow q)$ would be identical to those in the column for $p \wedge q$. We know this without actually making the table. In this case, the table would only be four lines long and easy enough to make. But Boolean algebra can be applied in cases where the number of propositional variables is too large for a truth table to be practical.

Let's do another example. Recall that a compound proposition is a tautology if it is true for all possible combinations of truth values of the propositional variables that it contains. But another way of saying the same thing is that P is a tautology if $P \equiv \mathbb{T}$. So, we can prove that a compound proposition, P, is a tautology by finding a chain of logical equivalences leading from P to $\mathbb{T}$. For example:

$$\begin{aligned}
&((p \vee q) \wedge \neg p) \rightarrow q \\
&\qquad \equiv (\neg((p \vee q) \wedge \neg p)) \vee q && \text{definition of } \rightarrow \\
&\qquad \equiv (\neg(p \vee q) \vee \neg(\neg p)) \vee q && \text{DeMorgan's Law, Theorem 1.2} \\
&\qquad \equiv (\neg(p \vee q) \vee p) \vee q && \text{Double Negation, Theorem 1.2} \\
&\qquad \equiv (\neg(p \vee q)) \vee (p \vee q) && \text{Associative Law for } \vee \\
&\qquad \equiv \mathbb{T} && \text{Law of Excluded Middle}
\end{aligned}$$

From this chain of equivalences, we can conclude that $((p \vee q) \wedge \neg p) \rightarrow q$ is a tautology.

Now, it takes some practice to look at an expression and see which rules can be applied to it; to see $(\neg(p \vee q)) \vee (p \vee q)$ as an application of the law of the excluded middle for example, you need to mentally substitute $(p \vee q)$ for p in the law as it is stated in Figure 1.2. Often, there are several rules that apply, and there are no definite guidelines about which one you should try. This is what makes algebra something of an art.

It is certainly not true that all possible rules of Boolean algebra are given in Figure 1.2. For one thing, there are many rules that are easy consequences of the rules that are listed there. For example, although the table asserts only that $\mathbb{F} \vee p \equiv p$, it is also true that $p \vee \mathbb{F} \equiv p$. This can be checked directly or by a simple calculation:

$$\begin{aligned}
p \vee \mathbb{F} &\equiv \mathbb{F} \vee p && \text{Commutative Law} \\
&\equiv p && \text{Identity Law as given in the table}
\end{aligned}$$

Additional rules can be obtained by applying the Commutative Law to other rules in the table, and we will use such rules freely in the future.

Another sort of easy extension can be applied to the Associative Law, $(p \vee q) \vee r \equiv p \vee (q \vee r)$. The law is stated for the $\vee$ operator applied to three terms, but it generalizes to four or more terms. For example

$$\begin{aligned}
&((p \vee q) \vee r) \vee s \\
&\qquad \equiv (p \vee q) \vee (r \vee s) && \text{by the Associative Law for three terms} \\
&\qquad \equiv p \vee (q \vee (r \vee s)) && \text{by the Associative Law for three terms}
\end{aligned}$$

We will, of course, often write this expression as $p \vee q \vee r \vee s$, with no parentheses at all, knowing that wherever we put the parentheses the value is the same.

One other thing that you should keep in mind is that rules can be applied in either direction. The Distributive Law, for example, allows you to distribute the p in $p \vee (q \wedge \neg p)$ to get $(p \vee q) \wedge (p \vee \neg p)$. But it can also be used in reverse to "factor out" a term, as when you start with $(q \vee (p \rightarrow q)) \wedge (q \vee (q \rightarrow p))$ and factor out the q to get $q \vee ((p \rightarrow q) \wedge (q \rightarrow p))$.

So far in this section, I have been working with the laws of Boolean algebra without saying much about what they mean or why they are reasonable. Of course, you can apply the laws in calculations without understanding them. But if you want to figure out *which* calculations to do, you need some understanding. Most of the laws are clear enough with a little thought. For example, if we already know that q is false, then $p \vee q$ will be true when p is true and false when p is false. That is, $p \vee \mathbb{F}$ has the same logical value as p. But that's just what the Identity Law for $\vee$ says. A few of the laws need more discussion.

The Law of the Excluded Middle, $p \vee \neg p \equiv \mathbb{T}$, says that given any proposition p, at least one of p or $\neg p$ must be true. Since $\neg p$ is true exactly when p is false, this is the same as saying that p must be either true or false. There is no middle ground.[7] The Law of Contradiction, $p \wedge \neg p \equiv \mathbb{F}$, says that it is not possible for *both* p and $\neg p$ to be true. Both of these rules are obvious.

The Distributive Laws cannot be called obvious, but a few examples can show that they are reasonable. Consider the statement, "This card is the ace of spades or clubs." Clearly, this is equivalent to "This card is the ace of spaces or this card is the ace of clubs." But this is just an example of the first distributive law! For, let a represent the proposition "This card is an ace," let s represent "This card is a spade," and let c represent "This card is a club." Then "This card is the ace of spades or clubs" can be translated into logic as $a \wedge (s \vee c)$, while "This card is the ace of spades or this card is the ace of clubs" becomes $(a \wedge s) \vee (a \wedge c)$. And the distributive law assures us that $a \wedge (s \vee c) \equiv (a \wedge s) \vee (a \wedge c)$. The second distributive law tells us, for example, that "This card is either a joker or is the ten of diamonds" is logically equivalent to "This card is either a joker or a ten, and it is either a

[7]In propositional logic, this is easily verified with a small truth table. But there is a surprising amount of argument about whether this law is valid in all situations. In the real world, there often seems to be a gray area between truth and falsity. Even in mathematics, there are some people who think there should be a third truth value, one that means something like "unknown" or "not proven." But the mathematicians who think this way tend to be considered a bit odd by most other mathematicians.

joker or a diamond." That is, $j \vee (t \wedge d) \equiv (j \vee t) \wedge (j \vee d)$. The distributive laws are powerful tools and you should keep them in mind whenever you are faced with a mixture of $\wedge$ and $\vee$ operators.

DeMorgan's Laws must also be less than obvious, since people often get them wrong. But they do make sense. When considering $\neg(p \wedge q)$, you should ask yourself, how can "p and q" *fail* to be true. It will fail to be true if either p is false *or* if q is false (or both). That is, $\neg(p \wedge q)$ is equivalent to $(\neg p) \vee (\neg q)$. Consider the sentence "A raven is large and black." If a bird is *not* large and black, then it is not a raven. But what exactly does it mean to be "*not (large and black)*"? How can you tell whether the assertion "not (large and black)" is true of something? This will be true if it is either not large *or* not black. (It doesn't have to be both—it could be large and white, it could be small and black.) Similarly, for "p or q" to fail to be true, *both* p and q must be false. That is, $\neg(p \vee q)$ is equivalent to $(\neg p) \wedge (\neg q)$. This is DeMorgan's second law.

Recalling that $p \to q$ is equivalent to $(\neg p) \vee q$, we can apply DeMorgan's law to obtain a formula for the negation an implication:

$$\begin{aligned} \neg(p \to q) &\equiv \neg((\neg p) \vee q) \\ &\equiv (\neg(\neg p)) \wedge (\neg q) \\ &\equiv p \wedge \neg q \end{aligned}$$

That is, $p \to q$ is false exactly when both p is true and q is false. For example, the negation of "If you have an ace, you win" is "You have an ace, and you don't win." Think of it this way: if you had an ace and you didn't win, then the statement "If you have an ace, you win" was not true.

Exercises

1. Construct truth tables to demonstrate the validity of each of the distributive laws.

2. Construct the following truth tables:
   a) Construct truth tables to demonstrate that $\neg(p \wedge q)$ is **not** logically equivalent to $(\neg p) \wedge (\neg q)$.
   b) Construct truth tables to demonstrate that $\neg(p \vee q)$ is **not** logically equivalent to $(\neg p) \vee (\neg q)$.
   c) Construct truth tables to demonstrate the validity of both DeMorgan's Laws.

3. Construct truth tables to demonstrate that $\neg(p \to q)$ is **not** logically equivalent to any of the following.

a) $(\neg p) \rightarrow (\neg q)$
b) $(\neg p) \rightarrow q$
c) $p \rightarrow (\neg q)$

Refer back to this section for a formula that **is** logically equivalent to $\neg(p \rightarrow q)$.

4. Is $\neg(p \leftrightarrow q)$ logically equivalent to $(\neg p) \leftrightarrow (\neg q)$?

5. In the algebra of numbers, there is a distributive law of multiplication over addition: $x(y+z) = xy + xz$. What would a distributive law of addition over multiplication look like? Is it a valid law in the algebra of numbers?

6. The distributive laws given in Figure 1.2 are sometimes called the *left* distributive laws. The ***right distributive laws*** say that $(p \vee q) \wedge r \equiv (p \wedge r) \vee (q \wedge r)$ and that $(p \wedge q) \vee r \equiv (p \vee r) \wedge (q \vee r)$. Show that the right distributive laws are also valid laws of Boolean algebra. (Note: In practice, both the left and the right distributive laws are referred to simply as the distributive laws, and both can be used freely in proofs.)

7. Show that $p \wedge (q \vee r \vee s) \equiv (p \wedge q) \vee (p \wedge r) \vee (p \wedge s)$ for any propositions p, q, r, and s. In words, we can say that conjunction distributes over a disjunction of three terms. (Recall that the $\wedge$ operator is called conjunction and $\vee$ is called disjunction.) Translate into logic and verify the fact that conjunction distributes over a disjunction of four terms. Argue that, in fact, conjunction distributes over a disjunction of any number of terms.

8. There are two additional basic laws of logic, involving the two expression $p \wedge \mathbb{F}$ and $p \vee \mathbb{T}$. What are the missing laws? Show that your answers are, in fact, laws.

9. For each of the following pairs of propositions, show that the two propositions are logically equivalent by finding a chain of equivalences from one to the other. State which definition or law of logic justifies each equivalence in the chain.

a) $p \wedge (q \wedge p), \quad p \wedge q$
b) $(\neg p) \rightarrow q, \quad p \vee q$
c) $(p \vee q) \wedge \neg q, \quad p \wedge \neg q$
d) $p \rightarrow (q \rightarrow r), \quad (p \wedge q) \rightarrow r$
e) $(p \rightarrow r) \wedge (q \rightarrow r), \quad (p \vee q) \rightarrow r$
f) $p \rightarrow (p \wedge q), \quad p \rightarrow q$

10. For each of the following compound propositions, find a simpler proposition that is logically equivalent. Try to find a proposition that is as simple as possible.

a) $(p \wedge q) \vee \neg q$
b) $\neg(p \vee q) \wedge p$
c) $p \rightarrow \neg p$
d) $\neg p \wedge (p \vee q)$
e) $(q \wedge p) \rightarrow q$
f) $(p \rightarrow q) \wedge (\neg p \rightarrow q)$

11. Express the negation of each of the following sentences in natural English:

a) It is sunny and cold.
b) I will have cake or I will have pie.
c) If today is Tuesday, this is Belgium.
d) If you pass the final exam, you pass the course.

12. Apply one of the laws of logic to each of the following sentences, and rewrite it as an equivalent sentence. State which law you are applying.

a) I will have coffee and cake or pie.
b) He has neither talent nor ambition.
c) You can have spam, or you can have spam.

1.3 Application: Logic Circuits

Computers have a reputation—not always deserved—for being "logical." But fundamentally, deep down, they are made of logic in a very real sense. The building blocks of computers are ***logic gates***, which are electronic components that compute the values of simple propositions such as $p \wedge q$ and $\neg p$. (Each gate is in turn built of even smaller electronic components called transistors, but this needn't concern us here.)

A wire in a computer can be in one of two states, which we can think of as being *on* and *off*. These two states can be naturally associated with the Boolean values $\mathbb{T}$ and $\mathbb{F}$. When a computer computes, the multitude of wires inside it are turned on and off in patterns that are determined by certain rules. The rules involved can be most naturally expressed in terms of logic. A simple rule might be, "turn wire C on whenever wire A is on and wire B is on." This rule can be implemented in hardware as an ***AND gate***. An AND gate is an electronic component with two input wires and one output wire, whose job is to turn its output on when both of its inputs are on and to turn its output off in any other case. If we associate "on" with $\mathbb{T}$ and "off" with $\mathbb{F}$, and if we give the names A and B to the inputs of the gate, then the gate computes the value of the logical expression $A \wedge B$. In effect, A is a proposition with the meaning "the first input is on," and B is a proposition with the meaning "the second input is on." The AND gate functions to ensure that the output is described by the proposition $A \wedge B$. That is, the output is on if and only if the first input is on and the second input is on.

An ***OR gate*** is an electronic component with two inputs and one output which turns its output on if either (or both) of its inputs is on. If the inputs are given names A and B, then the OR gate computes the logical value of $A \vee B$. A ***NOT gate*** has one input and one output, and it turns its output off when the input is on and on when the input is off. If the input is named A, then the NOT gate computes the value of $\neg A$.

Other types of logic gates are, of course, possible. Gates could be made to compute $A \rightarrow B$ or $A \oplus B$, for example. However, any computation that can be performed by logic gates can be done using only AND, OR, and NOT gates, as we will see below. (In practice, however, NAND gates and NOR gates, which compute the values of $\neg(A \wedge B)$ and $\neg(A \vee B)$ respectively,

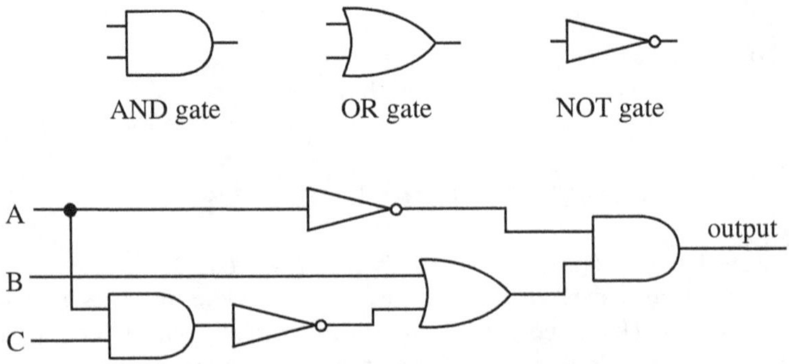

Figure 1.3: The standard symbols for the three basic logic gates, and a logic circuit that computes the value of the logical expression $(\neg A) \wedge (B \vee \neg(A \wedge C))$. The input wires to each logic gate are on the left, with the output wire on the right. Note that when wires cross each other in a diagram such as this, the wires don't actually intersect unless there is a black circle at the point where they cross.

are often used because they are easier to build from transistors than AND and OR gates.)

The three types of logic gates are represented by standard symbols, as shown in Figure 1.3. Since the inputs and outputs of logic gates are just wires carrying on/off signals, logic gates can be wired together by connecting outputs from some gates to inputs of other gates. The result is a ***logic circuit***. An example is also shown in Figure 1.3.

The logic circuit in the figure has three inputs, labeled A, B, and C. The circuit computes the value of the compound proposition $(\neg A) \wedge (B \vee \neg(A \wedge C))$. That is, when A represents the proposition "the input wire labeled A is on," and similarly for B and C, then the output of the circuit is on if and only if the value of the compound proposition $(\neg A) \wedge (B \vee \neg(A \wedge C))$ is true.

Given any compound proposition made from the operators $\wedge$, $\vee$, and $\neg$, it is possible to build a logic circuit that computes the value of that proposition. The proposition itself is a blueprint for the circuit. As noted in Section 1.1, every logical operator that we have encountered can be expressed in terms of $\wedge$, $\vee$, and $\neg$, so in fact every compound proposition that we know how to write can be computed by a logic circuit.

Given a proposition constructed from $\wedge$, $\vee$, and $\neg$ operators, it is easy to build a circuit to compute it. First, identify the main operator in the proposition—the one whose value will be computed *last*. Consider $(A \vee B) \wedge \neg(A \wedge B)$. This circuit has two input values, A and B, which are represented by wires coming into the circuit. The circuit has an output wire that represents the computed value of the proposition. The main operator in $(A \vee B) \wedge \neg(A \wedge B)$, is the first $\wedge$, which computes the value of the expression as a whole by combining the values of the subexpressions $A \vee B$ and $\neg(A \wedge B)$. This $\wedge$ operator corresponds to an AND gate in the circuit that computes the final output of the circuit.

Once the main operator has been identified and represented as a logic gate, you just have to build circuits to compute the input or inputs to that operator. In the example, the inputs to the main AND gate come from two subcircuits. One subcircuit computes the value of $A \vee B$ and the other computes the value of $\neg(A \wedge B)$. Building each subcircuit is a separate problem, but smaller than the problem you started with. Eventually, you'll come to a gate whose input comes directly from one of the input wires—A or B in this case—instead of from a subcircuit.

So, every compound proposition is computed by a logic circuit with one output wire. Is the reverse true? That is, given a logic circuit with one output, is there a proposition that expresses the value of the output in terms of the values of the inputs? Not quite. When you wire together some logic gates to make a circuit, there is nothing to stop you from introducing feedback loops. A feedback loop occurs when the output from a gate is connected—possibly through one or more intermediate gates—back to an input of the same gate. Figure 1.5 shows an example of a circuit with a feedback loop. Feedback loops cannot be described by compound propositions, basically because there is no place to start, no input to associate with a propositional variable. But feedback loops are the only thing that can go wrong. A logic circuit that does not contain any feedback loops is called a ***combinatorial logic circuit***. Every combinatorial logic circuit with just one output computes the value of some compound proposition. The propositional variables in the compound proposition are just names associated with the input wires of the circuit. (Of course, if the circuit has more than one output, you can simply use a different proposition for each output.)

The key to understanding why this is true is to note that each wire in the circuit—not just the final output wire—represents the value of some proposition. Furthermore, once you know which proposition is represented by each input wire to a gate, it's obvious what proposition is represented by the output: You just combine the input propositions with the appropriate $\wedge$, $\vee$, or $\neg$ operator, depending on what type of gate it is. To find the

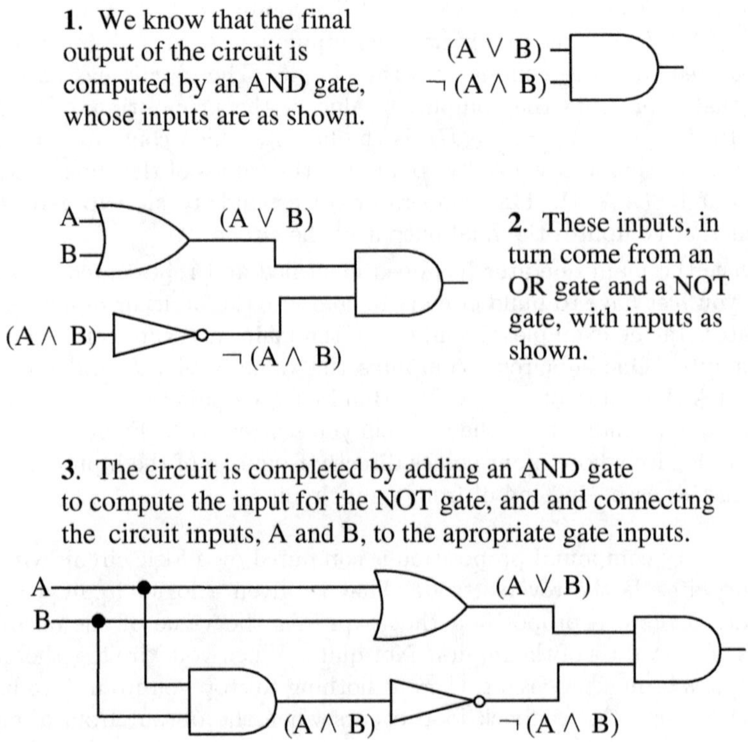

Figure 1.4: Stages in the construction of a circuit that computes the compound proposition $(A \vee B) \wedge \neg(A \wedge B)$.

proposition associated with the final output, you just have to start from the inputs and move through the circuit, labeling the output wire of each gate with the proposition that it represents. Figure 1.6 illustrates this process.

So, compound propositions correspond naturally with combinatorial logic circuits. But we have still not quite settled the question of just how powerful these circuits and propositions are. We've looked at a number of logical operators and noted that they can all be expressed in terms of $\wedge$, $\vee$, and $\neg$. But might there be other operators that cannot be so expressed? Equivalently, might there be other types of logic gates—possibly with some large number of inputs—whose computations cannot be duplicated with AND, OR, and NOT gates? Any logical operator or logic gate computes a

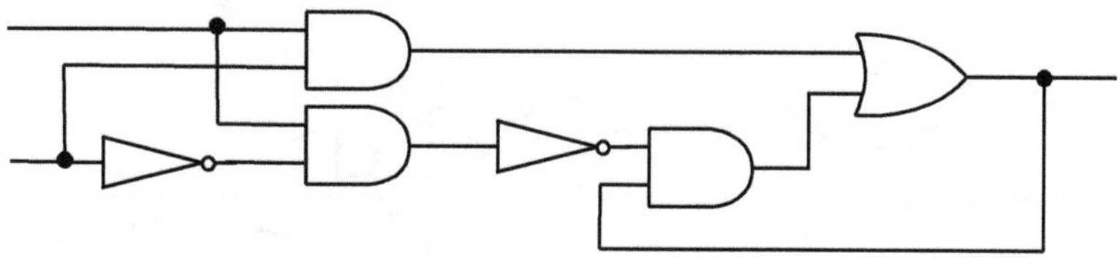

Figure 1.5: This circuit contains a feedback loop, so it is not a combinatorial logic circuit. The feedback loop includes the AND *gate and the* OR *gate on the right. This circuit does not compute the value of a compound proposition. This circuit does, however, play an important role in computer memories, since it can be used to store a logical value.*

value for each possible combination of logical values of its inputs. We could always make a truth table showing the output for each possible combination of inputs. As it turns out, given *any* such truth table, it is possible to find a proposition, containing only the $\wedge$, $\vee$, and $\neg$ operators, whose value for each combination of inputs is given precisely by that table.

To see why this is true, it is useful to introduce a particular type of compound proposition. Define a ***simple term*** to be either a propositional variable or the negation of a propositional variable. A conjunction of simple terms would then consist of one or more simple terms put together with $\wedge$ operators. (A "conjunction of one simple term" is just a single simple term by itself. This might not make grammatical sense, but it's the way mathematicians think.) Some examples of conjunctions of simple terms would be $p \wedge q$, p, $\neg q$, and $p \wedge \neg r \wedge \neg w \wedge s \wedge t$. Finally, we can take one or more such conjunctions and join them into a "disjunction of conjunctions of simple terms." This is the type of compound proposition we need. We can avoid some redundancy by assuming that no propositional variable occurs more than once in a single conjunction (since $p \wedge p$ can be replaced by p, and if p and $\neg p$ both occur in a conjunction, then the value of the conjuction is false, and it can be eliminated.) We can also assume that the same conjunction does not occur twice in the disjunction.

Definition 1.5. A compound proposition is said to be in ***disjunctive normal form***, or DNF, if it is a disjunction of conjunctions of simple terms, and if, furthermore, each propositional variable occurs at most once in each conjunction and each conjunction occurs at most once in the disjunction.

Using p, q, r, s, A, and B as propositional variables, here are a few

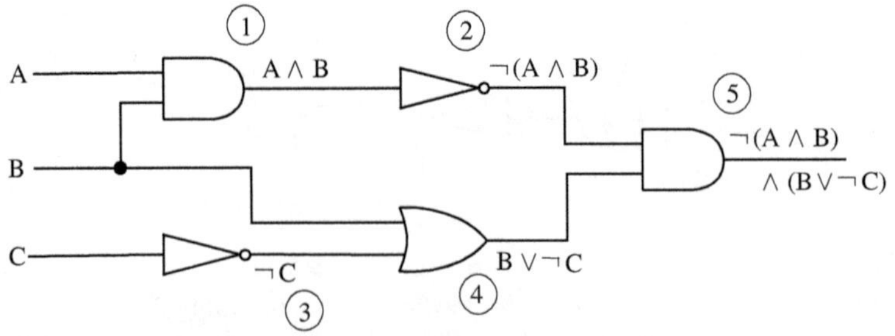

Figure 1.6: Finding the proposition whose value is computed by a combinatorial logic circuit. Each wire in the circuit is labeled with the proposition that it represents. The numbering of the labels shows one of the orders in which they can be associated with the wires. The circuit as a whole computes the value of $\neg(A \wedge B) \wedge (B \vee \neg C)$.

examples of propositions that are in disjunctive normal form:

$$(p \wedge q \wedge r) \vee (p \wedge \neg q \wedge r \wedge s) \vee (\neg p \wedge \neg q)$$
$$(p \wedge \neg q)$$
$$(A \wedge \neg B) \vee (\neg A \wedge B)$$
$$p \vee (\neg p \wedge q) \vee (\neg p \wedge \neg q \wedge r) \vee (\neg p \wedge \neg q \wedge \neg r \wedge w)$$

Propositions in DNF are just what we need to deal with input/output tables of the type that we have been discussing. Any such table can be computed by a proposition in disjunctive normal form. It follows that it is possible to build a circuit to compute that table using only AND, OR, and NOT gates.

Theorem 1.3. *Consider a table that lists a logical output value for every combination of values of several propositional variables. Assume that at least one of the output values is true. Then there is a proposition containing those variables such that the value of the proposition for each possible combination of the values of the variables is precisely the value specified in the table. It is possible to choose the proposition to be in disjunctive normal form.*

Proof. Consider any row in the table for which the output value is $\mathbb{T}$. Form a conjunction of simple terms as follows: For each variable, p, whose value is $\mathbb{T}$ in that row, include p itself in the conjunction; for each variable, q, whose value is $\mathbb{F}$ in the row, include $\neg q$ in the conjunction. The value of this conjunction is $\mathbb{T}$ for the combination of variable values given in that row of the table, since each of the terms in the conjuction is true for that combination of variables. Furthermore, for any *other* possible combination of variable values, the value of the conjunction will be $\mathbb{F}$, since at least one of the simple terms in the conjunction will be false.

Take the disjunction of all such conjunctions constructed in this way, for each row in the table where the output value is true. This disjunction has the value $\mathbb{T}$ if and only if one of the conjunctions that make it up has the value $\mathbb{T}$—and that is precisely when the output value specified by the table is $\mathbb{T}$. So, this disjunction of conjunctions satisfies the requirements of the theorem. □

As an example, consider the table in Figure 1.7. This table specifies a desired output value for each possible combination of values for the propositional variables p, q, and r. Look at the second row of the table, where the output value is true. According to the proof of the theorem, this row corresponds to the conjunction $(\neg p \wedge \neg q \wedge r)$. This conjunction is true when p is false, q is false, and r is true; in all other cases it is false, since in any other case at least one of the terms $\neg p$, $\neg q$, or r is false. The other two rows where the output is true give two more conjunctions. The three conjunctions are combined to produce the DNF proposition $(\neg p \wedge \neg q \wedge r) \vee (\neg p \wedge q \wedge r) \vee (p \wedge q \wedge r)$. This proposition computes all the output values specified in the table. Using this proposition as a blueprint, we get a logic circuit whose outputs match those given in the table.

Now, given any combinatorial logic circuit, there are many other circuits that have the same input/output behavior. When two circuits have the same input/output table, the compound propositions associated with the two circuits are logically equivalent. To put this another way, propositions that are logically equivalent produce circuits that have the same input/output behavior. As a practical matter, we will usually prefer the circuit that is simpler. The correspondence between circuits and propositions allows us to apply Boolean algebra to the simplification of circuits.

For example, consider the DNF proposition corresponding to the table in Figure 1.7. In $(\neg p \wedge \neg q \wedge r) \vee (\neg p \wedge q \wedge r) \vee (p \wedge q \wedge r)$, we can factor $(q \wedge r)$ from the last two terms, giving $(\neg p \wedge \neg q \wedge r) \vee ((\neg p \vee p) \wedge (q \wedge r))$. Since $\neg p \vee p \equiv \mathbb{T}$, and $\mathbb{T} \wedge Q \equiv Q$ for any proposition Q, this can be simplified to $(\neg p \wedge \neg q \wedge r) \vee (q \wedge r)$. Again, we can apply the distributive law to this to factor out an r, giving $((\neg p \wedge \neg q) \vee q) \wedge r)$. This compound proposition is logically equivalent to the one we started with, but implementing it in a circuit requires only five logic gates, instead of the ten required by the original proposition.[8]

[8]No, I didn't count wrong. There are eleven logical operators in the original expres-

p	q	r	output	
$\mathbb{F}$	$\mathbb{F}$	$\mathbb{F}$	$\mathbb{F}$	
$\mathbb{F}$	$\mathbb{F}$	$\mathbb{T}$	$\mathbb{T}$	$(\neg p \wedge \neg q \wedge r)$
$\mathbb{F}$	$\mathbb{T}$	$\mathbb{F}$	$\mathbb{F}$	
$\mathbb{F}$	$\mathbb{T}$	$\mathbb{T}$	$\mathbb{T}$	$(\neg p \wedge q \wedge r)$
$\mathbb{T}$	$\mathbb{F}$	$\mathbb{F}$	$\mathbb{F}$	
$\mathbb{T}$	$\mathbb{F}$	$\mathbb{T}$	$\mathbb{F}$	
$\mathbb{T}$	$\mathbb{T}$	$\mathbb{F}$	$\mathbb{F}$	
$\mathbb{T}$	$\mathbb{T}$	$\mathbb{T}$	$\mathbb{T}$	$p \wedge q \wedge r$

Figure 1.7: An input/output table specifying a desired output for each combination of values of the propositional variables p, q, and r. Each row where the output is $\mathbb{T}$ corresponds to a conjunction, shown next to that row in the table. The disjunction of these conjunctions is a proposition whose output values are precisely those specified by the table.

If you start with a circuit instead of a proposition, it is often possible to find the associated proposition, simplify it using Boolean algebra, and use the simplified proposition to build an equivalent circuit that is simpler than the original.

All this explains nicely the relationship between logic and circuits, but it doesn't explain why logic circuits should be used in computers in the first place. Part of the explanation is found in the fact that computers use binary numbers. A binary number is a string of zeros and ones. Binary numbers are easy to represent in an electronic device like a computer: Each position in the number corresponds to a wire. When the wire is on, it represents one; when the wire is off, it represents zero. When we are thinking in terms of logic, the same states of the wire represent true and false, but either representation is just an interpretation of the reality, which is a wire that is on or off. The question is whether the interpretation is fruitful.

Once wires are thought of as representing zeros and ones, we can build circuits to do computations with binary numbers. Which computations?

sion, but you can get by with ten gates in the circuit: Use a single NOT gate to compute $\neg p$, and connect the output of that gate to two different AND gates. Reusing the output of a logic gate is an obvious way to simplify circuits that does not correspond to any operation on propositions.

A	B	C	output
0	0	0	0
0	0	1	1
0	1	0	1
0	1	1	0
1	0	0	1
1	0	1	0
1	1	0	0
1	1	1	1

A	B	C	output
0	0	0	0
0	0	1	0
0	1	0	0
0	1	1	1
1	0	0	0
1	0	1	1
1	1	0	1
1	1	1	1

Figure 1.8: Input/output tables for the addition of three binary digits, A, B, and C.

Any that we want! If we know what the answer should be for each combination of inputs, then by Theorem 1.3 we can build a circuit to compute that answer. Of course, the procedure described in that theorem is only practical for small circuits, but small circuits can be used as building blocks to make all the calculating circuits in a computer.

For example, let's look at binary addition. To add two ordinary, decimal numbers, you line them up one on top of the other, and add the digits in each column. In each column, there might also be a carry from the previous column. To add up a column, you only need to remember a small number of rules, such as $7+6+1=14$ and $3+5+0=8$. For binary addition, it's even easier, since the only digits are 0 and 1. There are only eight rules:

$$\begin{array}{ll} 0+0+0=00 & 1+0+0=01 \\ 0+0+1=01 & 1+0+1=10 \\ 0+1+0=01 & 1+1+0=10 \\ 0+1+1=10 & 1+1+1=11 \end{array}$$

Here, I've written each sum using two digits. In a multi-column addition, one of these digits is carried over to the next column. Here, we have a calculation that has three inputs and two outputs. We can make an input/output table for each of the two outputs. The tables are shown in Figure 1.8. We know that these tables can be implemented as combinatorial circuits, so we know that circuits can add binary numbers. To add multi-digit binary numbers, we just need one copy of the basic addition circuit for each column in the sum.

Exercises

1. Using only AND, OR, and NOT gates, draw circuits that compute the value of each of the propositions $A \rightarrow B$, $A \leftrightarrow B$, and $A \oplus B$.

2. For each of the following propositions, find a combinatorial logic circuit that computes that proposition:

a) $A \wedge (B \vee \neg C)$ **b)** $(p \wedge q) \wedge \neg(p \wedge \neg q)$
c) $(p \vee q \vee r) \wedge (\neg p \vee \neg q \vee \neg r)$ **d)** $\neg(A \wedge (B \vee C)) \vee (B \wedge \neg A)$

3. Find the compound proposition computed by each of the following circuits:

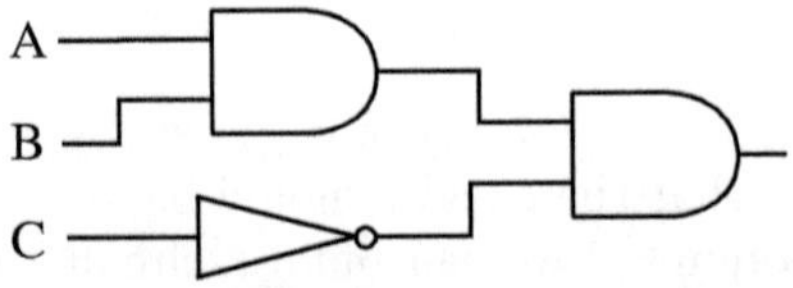

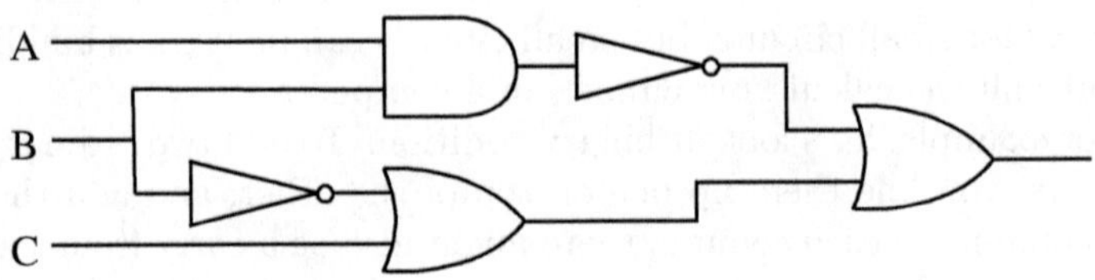

4. This section describes a method for finding the compound proposition computed by any combinatorial logic circuit. This method fails if you try to apply it to a circuit that contains a feedback loop. What goes wrong? Give an example.

5. Show that every compound proposition which is not a contradiction is equivalent to a proposition in disjunctive normal form. (Note: We can eliminate the restriction that the compound proposition is not a contradiction by agreeing that "$\mathbb{F}$" counts as a proposition in disjunctive normal form. $\mathbb{F}$ is logically equivalent to any contradiction.)

6. A proposition in ***conjunctive normal form*** (CNF) is a conjunction of disjunctions of simple terms (with the proviso, as in the definition of DNF that a single item counts as a conjunction or disjunction). Show that every compound proposition which is not a tautology is logically equivalent to a compound proposition in conjunctive normal form. (Hint: What happens if you take the negation of a DNF proposition and apply DeMorgan's Laws?)

7. Use the laws of Boolean algebra to simplify each of the following circuits:

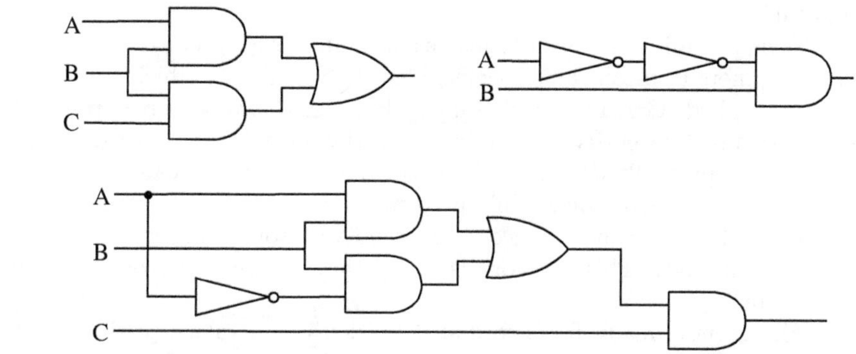

8. Design circuits to implement the input/output tables for addition, as given in Figure 1.8. Try to make your circuits as simple as possible. (The circuits that are used in real computers for this purpose are more simplified than the ones you will probably come up with, but the general approach of using logic to design computer circuits is valid.)

1.4 Predicates and Quantifiers

In propositional logic, we can let p stand for "Roses are red" and q stand for "Violets are blue." Then $p \wedge q$ will stand for "Roses are red and violets are blue." But we lose a lot in the translation into logic. Since propositional logic only deals with truth values, there's nothing we can do with p and q in propositional logic that has anything to do with roses, violets, or color. To apply logic to such things, we need ***predicates***. The type of logic that uses predicates is called ***predicate logic***, or, when the emphasis is on manipulating and reasoning with predicates, ***predicate calculus***.

A predicate is a kind of incomplete proposition, which becomes a proposition when it is applied to some entity (or, as we'll see later, to several entities). In the proposition "the rose is red," the predicate is *is red.* By itself, "is red" is not a proposition. Think of it as having an empty slot, that needs to be filled in to make a proposition: "— is red." In the proposition "the rose is red," the slot is filled by the entity "the rose," but it could just as well be filled by other entities: "the barn is red"; "the wine is red"; "the banana is red." Each of these propositions uses the same predicate, but they are different propositions and they can have different truth values.

If P is a predicate and a is an entity, then $P(a)$ stands for the proposition that is formed when P is applied to a. If P represents "is red" and a stands for "the rose," then $P(a)$ is "the rose is red." If M is the predicate "is mortal" and s is "Socrates," then $M(s)$ is the proposition "Socrates is mortal."

Now, you might be asking, just what is an *entity* anyway? I am using the term here to mean some specific, identifiable thing to which a predicate can be applied. Generally, it doesn't make sense to apply a given predicate to every possible entity, but only to entities in a certain category. For example, it probably doesn't make sense to apply the predicate "is mortal" to your living room sofa. This predicate only applies to entities in the category of living things, since there is no way something can be mortal unless it is alive. This category is called the domain of discourse for the predicate.[9]

We are now ready for a formal definition of one-place predicates. A one-place predicate, like all the examples we have seen so far, has a single slot which can be filled in with one entity:

Definition 1.6. A ***one-place predicate*** associates a proposition with each entity in some collection of entities. This collection is called the ***domain of discourse*** for the predicate. If P is a predicate and a is an entity in the domain of discourse for P, then $P(a)$ denotes the proposition that is associated with a by P. We say that $P(a)$ is the result of ***applying*** P to a.

We can obviously extend this to predicates that can be applied to two or more entities. In the proposition "John loves Mary," *loves* is a two-place predicate. Besides John and Mary, it could be applied to other pairs of entities: "John loves Jane," "Bill loves Mary," "John loves Bill," "John loves John." If Q is a two-place predicate, then $Q(a, b)$ denotes the proposition that is obtained when Q is applied to the entities a and b. Note that each of the "slots" in a two-place predicate can have its own domain of discourse. For example, if Q represents the predicate "owns," then $Q(a, b)$ will only make sense when a is a person and b is an inanimate object. An example of a three-place predicate is "a gave b to c," and a four-place predicate would be "a bought b from c for d dollars." But keep in mind that not every predicate has to correspond to an English sentence.

When predicates are applied to entities, the results are propositions, and all the operators of propositional logic can be applied to these propositions just as they can to any propositions. Let R be the predicate "is red," and let L be the two-place predicate "loves." If a, b, j, m, and b are entities belonging to the appropriate categories, then we can form compound

[9]In the language of set theory, which will be introduced in the next chapter, we would say that a domain of discourse is a set, U, and a predicate is a function from U to the set of truth values. The definition should be clear enough without the formal language of set theory, and in fact you should think of this definition—and many others—as motivation for that language.

propositions such as:

$R(a) \wedge R(b)$	a is red and b is red
$\neg R(a)$	a is not red
$L(j,m) \wedge \neg L(m,j)$	j loves m, and m does not love j
$L(j,m) \rightarrow L(b,m)$	if j loves m then b loves m
$R(a) \leftrightarrow L(j,j)$	a is red if and only if j loves j

Let's go back to the proposition with which we started this section: "Roses are red." This sentence is more difficult to handle than it might appear. We still can't express it properly in logic. The problem is that this proposition is not saying something about some particular entity. It really says that *all* roses are red (which happens to be a false statement, but that's what it means). Predicates can only be applied to individual entities.

Many other sentences raise similar difficulties: "All persons are mortal." "Some roses are red, but no roses are black." "All math courses are interesting." "Every prime number greater than two is odd." Words like *all*, *no*, *some*, and *every* are called ***quantifiers***. We need to be able to express similar concepts in logic.

Suppose that P is a predicate, and we want to express the proposition that P is true when applied to any entity in the domain of discourse. That is, we want to say "for any entity x in the domain of discourse, $P(x)$ is true." In predicate logic, we write this in symbols as $\forall x(P(x))$. The $\forall$ symbol, which looks like an upside-down A, is usually read "for all," so that $\forall x(P(x))$ is read as "for all x, $P(x)$." (It is understood that this means for all x in the domain of discourse for P.) For example, if R is the predicate "is red" and the domain of discourse consists of all roses, then $\forall x(R(x))$ expresses the proposition "All roses are red." Note that the same proposition could be expressed in English as "Every rose is red" or "Any rose is red."

Now, suppose we want to say that a predicate, P, is true for *some* entity in its domain of discourse. This is expressed in predicate logic as $\exists x(P(x))$. The $\exists$ symbol, which looks like a backwards E, is usually read "there exists," but a more exact reading would be "there is at least one." Thus, $\exists x(P(x))$ is read as "There exists an x such that $P(x)$," and it means "there is at least one x in the domain of discourse for P for which $P(x)$ is true." If, once again, R stands for "is red" and the domain of discourse is "roses," then $\exists x(R(x))$ could be expressed in English as "There is a red rose" or "At least one rose is red" or "Some rose is red." It might also be expressed as "Some roses are red," but the plural is a bit misleading since $\exists x(R(x))$ is true even if there is only one red rose. We can now give the formal definitions:

Definition 1.7. Suppose that P is a one-place predicate. Then $\forall x(P(x))$ is a proposition, which is true if and only if $P(a)$ is true for every entity a in the domain of discourse for P. And $\exists x(P(x))$ is a proposition which is true if and only if there is at least one entity, a, in the domain of discourse for P for which $P(a)$ is true. The $\forall$ symbol is called the ***universal quantifier***, and $\exists$ is called the ***existential quantifier***.

The x in $\forall x(P(x))$ and $\exists x(P(x))$ is a variable. (More precisely, it is an *entity* variable, since its value can only be an entity.) Note that a plain $P(x)$—without the $\forall x$ or $\exists x$—is not a proposition. $P(x)$ is neither true nor false because x is not some particular entity, but just a placeholder in a slot that can be filled in with an entity. $P(x)$ would stand for something like the statement "x is red," which is not really a statement in English at all. But it becomes a statement when the x is replaced by some particular entity, such as "the rose." Similarly, $P(x)$ becomes a proposition if some entity a is substituted for the x, giving $P(a)$.[10]

An ***open statement*** is an expression that contains one or more entity variables, which becomes a proposition when entities are substituted for the variables. (An open statement has open "slots" that need to be filled in.) $P(x)$ and "x is red" are examples of open statements that contain one variable. If L is a two-place predicate and x and y are variables, then $L(x, y)$ is an open statement containing two variables. An example in English would be "x loves y." The variables in an open statement are called ***free variables***. An open statement that contains x as a free variable can be quantified with $\forall x$ or $\exists x$. The variable x is then said to be **bound**. For example, x is free in $P(x)$ and is bound in $\forall x(P(x))$ and $\exists x(P(x))$. The free variable y in $L(x, y)$ becomes bound in $\forall y(L(x, y))$ and in $\exists y(L(x, y))$.

Note that $\forall y(L(x, y))$ is still an open statement, since it contains x as a free variable. Therefore, it is possible to apply the quantifier $\forall x$ or $\exists x$ to $\forall y(L(x, y))$, giving $\forall x\big(\forall y(L(x, y))\big)$ and $\exists x\big(\forall y(L(x, y))\big)$. Since all the variables are bound in these expressions, they are propositions. If $L(x, y)$ represents "x loves y," then $\forall y(L(x, y))$ is something like "x loves everyone," and $\exists x\big(\forall y(L(x, y))\big)$ is the proposition, "There is someone who loves everyone." Of course, we could also have started with $\exists x(L(x, y))$: "There is someone who loves y." Applying $\forall y$ to this gives $\forall y\big(\exists x(L(x, y))\big)$, which means "For every person, there is someone who loves that person." Note

[10] There is certainly room for confusion about names here. In this discussion, x is a variable and a is an entity. But that's only because I said so. Any letter could be used in either role, and you have to pay attention to the context to figure out what is going on. Usually, x, y, and z will be variables.

in particular that $\exists x\big(\forall y(L(x,y))\big)$ and $\forall y\big(\exists x(L(x,y))\big)$ do *not* mean the same thing. Altogether, there are eight different propositions that can be obtained from $L(x,y)$ by applying quantifiers, with six distinct meanings among them.

(From now on, I will leave out parentheses when there is no ambiguity. For example, I will write $\forall x\, P(x)$ instead of $\forall x(P(x))$ and $\exists x\, \exists y\, L(x,y)$ instead of $\exists x\big(\exists y(L(x,y))\big)$. Furthermore, I will sometimes give predicates and entities names that are complete words instead of just letters, as in $\mathit{Red}(x)$ and $\mathit{Loves}(\mathit{john},\mathit{mary})$. This might help to make examples more readable.)

In predicate logic, the operators and laws of Boolean algebra still apply. For example, if P and Q are one-place predicates and a is an entity in the domain of discourse, then $P(a) \rightarrow Q(a)$ is a proposition, and it is logically equivalent to $\neg P(a) \vee Q(a)$. Furthermore, if x is a variable, then $P(x) \rightarrow Q(x)$ is an open statement, and $\forall x(P(x) \rightarrow Q(x))$ is a proposition. So are $P(a) \wedge (\exists x\, Q(x))$ and $(\forall x\, P(x)) \rightarrow (\exists x P(x))$. Obviously, predicate logic can be very expressive. Unfortunately, the translation between predicate logic and English sentences is not always obvious.

Let's look one more time at the proposition "Roses are red." If the domain of discourse consists of roses, this translates into predicate logic as $\forall x\, \mathit{Red}(x)$. However, the sentence makes more sense if the domain of discourse is larger—for example if it consists of all flowers. Then "Roses are red" has to be read as "All flowers which are roses are red," or "For any flower, if that flower is a rose, then it is red." The last form translates directly into logic as $\forall x\big(\mathit{Rose}(x) \rightarrow \mathit{Red}(x)\big)$. Suppose we want to say that all red roses are pretty. The phrase "red rose" is saying both that the flower is a rose and that it is red, and it must be translated as a conjunction, $\mathit{Rose}(x) \wedge \mathit{Red}(x)$. So, "All red roses are pretty" can be rendered as $\forall x\big((\mathit{Rose}(x) \wedge \mathit{Red}(x)) \rightarrow \mathit{Pretty}(x)\big)$.

Here are a few more examples of translations from predicate logic to English. Let $H(x)$ represent "x is happy," let $C(y)$ represent "y is a computer," and let $O(x,y)$ represent "x owns y." (The domain of discourse for x consists of people, and the domain for y consists of inanimate objects.) Then we have the following translations:

- Jack owns a computer: $\exists x\big(O(\mathit{jack},x) \wedge C(x)\big)$. (That is, there is at least one thing such that Jack owns that thing and that thing is a computer.)
- Everything Jack owns is a computer: $\forall x\big(O(\mathit{jack},x) \rightarrow C(x)\big)$.

- If Jack owns a computer, then he's happy:
  $\big(\exists y(O(jack, y) \wedge C(y))\big) \rightarrow H(jack)$.
- Everyone who owns a computer is happy:
  $\forall x\big(\,\big(\exists y(O(x, y) \wedge C(y)\big) \rightarrow H(x))\,\big)$.
- Everyone owns a computer: $\forall x\, \exists y\big(C(y) \wedge O(x, y)\big)$. (Note that this allows each person to own a different computer. The proposition $\exists y\, \forall x\big(C(y) \wedge O(x, y)\big)$ would mean that there is a single computer which is owned by everyone.)
- Everyone is happy: $\forall x H(x)$.
- Everyone is unhappy: $\forall x(\neg H(x))$.
- Someone is unhappy: $\exists x(\neg H(x))$.
- At least two people are happy: $\exists x \exists y\big(H(x) \wedge H(y) \wedge (x \neq y)\big)$. (The stipulation that $x \neq y$ is necessary because two different variables can refer to the same entity. The proposition $\exists x \exists y(H(x) \wedge H(y))$ is true even if there is only one happy person.)
- There is exactly one happy person:
  $\big(\exists x H(x))\big) \wedge \big(\forall y \forall z((H(y) \wedge H(z)) \rightarrow (y = z))\big)$. (The first part of this conjunction says that there is at least one happy person. The second part says that if y and z are both happy people, then they are actually the same person. That is, it's not possible to find two *different* people who are happy.)

To calculate in predicate logic, we need a notion of logical equivalence. Clearly, there are pairs of propositions in predicate logic that mean the same thing. Consider the propositions $\neg(\forall x H(x))$ and $\exists x(\neg H(x))$, where $H(x)$ represents "x is happy." The first of these propositions means "Not everyone is happy," and the second means "Someone is not happy." These statements have the same truth value: If not everyone is happy, then someone is unhappy and vice versa. But logical equivalence is much stronger than just having the same truth value. In propositional logic, logical equivalence is defined in terms of propositional variables: two compound propositions are logically equivalent if they have the same truth values for all possible truth values of the propositional variables they contain. In predicate logic, two formulas are logically equivalent if they have the same truth value for all possible predicates.

Consider $\neg(\forall x P(x))$ and $\exists x(\neg P(x))$. These formulas make sense for any predicate P, and for any predicate P they have the same truth value. Unfortunately, we can't—as we did in propositional logic—just check this

$\neg(\forall x P(x)) \equiv \exists x(\neg P(x))$
$\neg(\exists x P(x)) \equiv \forall x(\neg P(x))$
$\forall x \forall y Q(x,y) \equiv \forall y \forall x Q(x,y)$
$\exists x \exists y Q(x,y) \equiv \exists y \exists x Q(x,y)$

Figure 1.9: Four important rules of predicate logic. P can be any one-place predicate, and Q can be any two-place predicate. The first two rules are called DeMorgan's Laws for predicate logic.

fact with a truth table: there are no subpropositions, connected by $\wedge$, $\vee$, etc, out of which to build a table. So, let's reason it out: To say $\neg(\forall x P(x))$ is true is just to say that it is not the case that $P(x)$ is true for all possible entities x. So, there must be some entity a for which $P(a)$ is false. Since $P(a)$ is false, $\neg P(a)$ is true. But saying that there is an a for which $\neg P(a)$ is true is just saying that $\exists x(\neg P(x))$ is true. So, the truth of $\neg(\forall x P(x))$ implies the truth of $\exists x(\neg P(x))$. On the other hand, if $\neg(\forall x P(x))$ is false, then $\forall x P(x)$ is true. Since $P(x)$ is true for every x, $\neg P(x)$ is false for every x; that is, there is no entity a for which the statement $\neg P(a)$ is true. But this just means that the statement $\exists x(\neg P(x))$ is false. In any case, then, the truth values of $\neg(\forall x P(x))$ and $\exists x(\neg P(x))$ are the same. Since this is true for any predicate P, we will say that these two formulas are logically equivalent and write $\neg(\forall x P(x)) \equiv \exists x(\neg P(x))$.

A similar argument would show that $\neg(\exists x P(x)) \equiv \forall x(\neg P(x))$. These two equivalences, which explicate the relation between negation and quantification, are known as DeMorgan's Laws for predicate logic. (They are closely related to DeMorgan's Laws for propositional logic; see the exercises.) These laws can be used to help simplify expressions. For example,

$$\begin{aligned} \neg\, \forall y(R(y) \vee Q(y)) &\equiv \exists y(\neg(R(y) \vee Q(y))) \\ &\equiv \exists y((\neg R(y)) \wedge (\neg Q(y)) \end{aligned}$$

It might not be clear exactly why this qualifies as a "simplification," but it's generally considered simpler to have the negation operator applied to basic propositions such as $R(y)$, rather than to quantified expressions such as $\forall y(R(y) \vee Q(y))$. For a more complicated example:

$$\begin{aligned}
&\neg\, \exists x\big(P(x) \wedge (\forall y(Q(y) \rightarrow Q(x)))\big)\\
&\qquad \equiv \forall x\big(\neg(P(x) \wedge (\forall y(Q(y) \rightarrow Q(x))))\big)\\
&\qquad \equiv \forall x\big((\neg P(x)) \vee (\neg\forall y(Q(y) \rightarrow Q(x)))\big)\\
&\qquad \equiv \forall x\big((\neg P(x)) \vee (\exists y(\neg(Q(y) \rightarrow Q(x))))\big)\\
&\qquad \equiv \forall x\big((\neg P(x)) \vee (\exists y(\neg(\neg Q(y) \vee Q(x))))\big)\\
&\qquad \equiv \forall x\big((\neg P(x)) \vee (\exists y(\neg\neg Q(y) \wedge \neg Q(x)))\big)\\
&\qquad \equiv \forall x\big((\neg P(x)) \vee (\exists y(Q(y) \wedge \neg Q(x)))\big)
\end{aligned}$$

DeMorgan's Laws are listed in Figure 1.9 along with two other laws of predicate logic. The other laws allow you to interchange the order of the variables when two quantifiers of the same type (both $\exists$ or $\forall$) occur together.

To define logical equivalence in predicate logic more formally, we need to talk about formulas that contain predicate variables, that is, variables that act as place-holders for arbitrary predicates in the same way that propositional variables are place-holders for propositions and entity variables are place-holders for entities. With this in mind, we can define logical equivalence and the closely related concept of tautology for predicate logic.

Definition 1.8. Let $\mathcal{P}$ be a formula of predicate logic which contains one or more predicate variables. $\mathcal{P}$ is said to be a ***tautology*** if it is true whenever all the predicate variables that it contains are replaced by actual predicates. Two formulas $\mathcal{P}$ and $\mathcal{Q}$ are said to be ***logically equivalent*** if $\mathcal{P} \leftrightarrow \mathcal{Q}$ is a tautology, that is if $\mathcal{P}$ and $\mathcal{Q}$ always have the same truth value when the predicate variables they contain are replaced by actual predicates. The notation $\mathcal{P} \equiv \mathcal{Q}$ asserts that $\mathcal{P}$ is logically equivalent to $\mathcal{Q}$.

Exercises

1. Simplify each of the following propositions. In your answer, the $\neg$ operator should be applied only to individual predicates.
   - **a)** $\neg\,\forall x(\neg P(x))$
   - **b)** $\neg\,\exists x(P(x) \wedge Q(x))$
   - **c)** $\neg\,\forall z(P(z) \rightarrow Q(z))$
   - **d)** $\neg\big((\forall x P(x)) \wedge \forall y(Q(y))\big)$
   - **e)** $\neg\,\forall x \exists y P(x, y)$
   - **f)** $\neg\,\exists x(R(x) \wedge \forall y S(x, y))$
   - **g)** $\neg\,\exists y(P(y) \leftrightarrow Q(y))$
   - **h)** $\neg\big(\forall x(P(x) \rightarrow (\exists y Q(x, y)))\big)$
2. Give a careful argument to show that the second of DeMorgan's laws for predicate calculus, $\neg(\forall x P(x)) \equiv \exists x(\neg P(x))$, is valid.

3. Find the negation of each of the following propositions. Simplify the result; in your answer, the $\neg$ operator should be applied only to individual predicates.

a) $\neg\exists n(\forall s C(s,n))$

b) $\neg\exists n(\forall s(L(s,n) \rightarrow P(s)))$

c) $\neg\exists n(\forall s(L(s,n) \rightarrow (\exists x\exists y\exists z Q(x,y,z))))$.

d) $\neg\exists n(\forall s(L(s,n) \rightarrow (\exists x\exists y\exists z(s = xyz \wedge R(x,y) \wedge T(y) \wedge U(x,y,z))))$.

4. Suppose that the domain of discourse for a predicate P contains only two entities. Show that $\forall x P(x)$ is equivalent to a conjunction of two simple propositions, and $\exists x P(x)$ is equivalent to a disjunction. Show that in this case, DeMorgan's Laws for propositional logic and DeMorgan's Laws for predicate logic actually say exactly the same thing. Extend the results to a domain of discourse that contains exactly three entities.

5. Let $H(x)$ stand for "x is happy," where the domain of discourse consists of people. Express the proposition "There are exactly three happy people" in predicate logic.

6. Let $T(x,y)$ stand for "x has taken y," where the domain of discourse for x consists of students and the domain of discourse for y consists of math courses (at your school). Translate each of the following propositions into an unambiguous English sentence:

a) $\forall x\,\forall y\,T(x,y)$ **b)** $\forall x\,\exists y\,T(x,y)$ **c)** $\forall y\,\exists x\,T(x,y)$

d) $\exists x\,\exists y\,T(x,y)$ **e)** $\exists x\,\forall y\,T(x,y)$ **f)** $\exists y\,\forall x\,T(x,y)$

7. Let $F(x,t)$ stand for "You can fool person x at time t." Translate the following sentence into predicate logic: "You can fool some of the people all of the time, and you can fool all of the people some of the time, but you can't fool all of the people all of the time."

8. Translate each of the following sentences into a proposition using predicate logic. Make up any predicates you need. State what each predicate means and what its domain of discourse is.

a) All crows are black.

b) Any white bird is not a crow.

c) Not all politicians are honest.

d) All green elephants have purple feet.

e) There is no one who does not like pizza.

f) Anyone who passes the final exam will pass the course.

g) If x is any positive number, then there is a number y such that $y^2 = x$.

9. The sentence "Someone has the answer to every question" is ambiguous. Give two translations of this sentence into predicate logic, and explain the difference in meaning.

10. The sentence "Jane is looking for a dog" is ambiguous. One meaning is that there is some particular dog—maybe the one she lost—that Jane is looking for. The other meaning is that Jane is looking for any old dog—maybe because she wants to buy one. Express the first meaning in predicate logic. Explain why the second meaning is *not* expressed by $\forall x(Dog(x) \rightarrow LooksFor(jane,x))$. In fact, the second meaning cannot be expressed in predicate logic. Philosophers of language spend a lot of time thinking about things like this. They are especially fond of the sentence "Jane is looking for a unicorn," which is not ambiguous when applied to the real world. Why is that?

1.5 Deduction

Logic can be applied to draw conclusions from a set of premises. A premise is just a proposition that is known to be true or that has been accepted to be true for the sake of argument, and a conclusion is a proposition that can be deduced logically from the premises. The idea is that if you believe that the premises are true, then logic forces you to accept that the conclusion is true. An "argument" is a claim that a certain conclusion follows from a given set of premises. Here is an argument laid out in a traditional format:

If today is Tuesday, then this is Belgium
Today is Tuesday
___
∴ This is Belgium

The premises of the argument are shown above the line, and the conclusion below. The symbol ∴ is read "therefore." The claim is that the conclusion, "This is Belgium," can be deduced logically from the two premises, "If today is Tuesday, then this is Belgium" and "Today is Tuesday." In fact, this claim is true. Logic forces you to accept this argument. Why is that?

Let p stand for the proposition "Today is Tuesday," and let q stand for the proposition "This is Belgium." Then the above argument has the form

$$\begin{array}{l} p \rightarrow q \\ \underline{p \qquad} \\ \therefore\ q \end{array}$$

Now, for *any* propositions p and q—not just the ones in this particular argument—if $p \rightarrow q$ is true and p is true, then q must also be true. This is easy to check in a truth table:

p	q	$p \rightarrow q$
false	false	true
false	true	true
true	false	false
true	true	true

The only case where both $p \rightarrow q$ and p are true is on the last line of the table, and in this case, q is also true. If you believe $p \rightarrow q$ and p, you have no logical choice but to believe q. This applies no matter what p and q represent. For example, if you believe "If Jill is breathing, then Jill pays taxes," and you believe that "Jill is breathing," logic forces you to believe that "Jill pays taxes." Note that we can't say for sure that the conclusion is true, only that *if* the premises are true, *then* the conclusion must be true.

This fact can be rephrased by saying that $\big((p \to q) \land p\big) \to q$ is a tautology. More generally, for any compound propositions $\mathcal{P}$ and $\mathcal{Q}$, saying "$\mathcal{P} \to \mathcal{Q}$ is a tautology" is the same as saying that "in all cases where $\mathcal{P}$ is true, $\mathcal{Q}$ is also true".[11] We will use the notation $\mathcal{P} \Longrightarrow \mathcal{Q}$ to mean that $\mathcal{P} \to \mathcal{Q}$ is a tautology. Think of $\mathcal{P}$ as being the premise of an argument or the conjunction of several premises. To say $\mathcal{P} \Longrightarrow \mathcal{Q}$ is to say that $\mathcal{Q}$ follows logically from $\mathcal{P}$. We will use the same notation in both propositional logic and predicate logic. (Note that the relation of $\Longrightarrow$ to $\to$ is the same as the relation of $\equiv$ to $\leftrightarrow$.)

Definition 1.9. Let $\mathcal{P}$ and $\mathcal{Q}$ be any formulas in either propositional logic or predicate logic. The notation $\mathcal{P} \Longrightarrow \mathcal{Q}$ is used to mean that $\mathcal{P} \to \mathcal{Q}$ is a tautology. That is, in all cases where $\mathcal{P}$ is true, $\mathcal{Q}$ is also true. We then say that $\mathcal{Q}$ can be ***logically deduced*** from $\mathcal{P}$ or that $\mathcal{P}$ ***logically implies*** $\mathcal{Q}$.

An argument in which the conclusion follows logically from the premises is said to be a ***valid argument***. To test whether an argument is valid, you have to replace the particular propositions or predicates that it contains with variables, and then test whether the conjunction of the premises logically implies the conclusion. We have seen that any argument of the form

$$\begin{array}{l} p \to q \\ \underline{p \qquad\quad} \\ \therefore\ q \end{array}$$

is valid, since $\big((p \to q) \land p\big) \to q$ is a tautology. This rule of deduction is called ***modus ponens***. It plays a central role in logic. Another, closely related rule is ***modus tollens***, which applies to arguments of the form

$$\begin{array}{l} p \to q \\ \underline{\neg q \qquad\ } \\ \therefore\ \neg p \end{array}$$

[11]Here, "in all cases" means for all combinations of truth values of the propositional variables in $\mathcal{P}$ and $\mathcal{Q}$. Saying $\mathcal{P} \to \mathcal{Q}$ is a tautology means it is true in all cases. But by definition of $\to$, it is automatically true in cases where $\mathcal{P}$ is false. In cases where $\mathcal{P}$ is true, $\mathcal{P} \to \mathcal{Q}$ will be true if and only if $\mathcal{Q}$ is true.

To verify that this is a valid argument, just check that $\big((p \to q) \wedge \neg q\big) \Longrightarrow \neg p$, that is, that $\big((p \to q) \wedge \neg q\big) \to \neg p$ is a tautology. As an example, the following argument has the form of *modus tollens* and is therefore a valid argument:

If Keanu Reeves is a good actor, then I'm the king of France
I am not the king of France

∴ Keanu Reeves in not a good actor

You should note carefully that the validity of this argument has nothing to do with whether or not Keanu Reeves can act well. The argument forces you to accept the conclusion *only if* you accept the premises. You can logically believe that the conclusion is false, as long as you believe that at least one of the premises is false.

Another named rule of deduction is the ***Law of Syllogism***, which has the form

$$\begin{array}{l} p \to q \\ \underline{q \to r} \\ \therefore\ p \to r \end{array}$$

For example:

If you study hard, you do well in school
If you do well in school, you get a good job

∴ If you study hard, you get a good job

There are many other rules. Here are a few that might prove useful. Some of them might look trivial, but don't underestimate the power of a simple rule when it is combined with other rules.

$$\begin{array}{l} p \vee q \\ \underline{\neg p} \\ \therefore\ q \end{array} \qquad \begin{array}{l} p \\ \underline{q} \\ \therefore\ p \wedge q \end{array} \qquad \begin{array}{l} \underline{p \wedge q} \\ \therefore\ p \end{array} \qquad \begin{array}{l} \underline{p} \\ \therefore\ p \vee q \end{array}$$

Logical deduction is related to logical equivalence. We defined $\mathcal{P}$ and $\mathcal{Q}$ to be logically equivalent if $\mathcal{P} \leftrightarrow \mathcal{Q}$ is a tautology. Since $\mathcal{P} \leftrightarrow \mathcal{Q}$ is equivalent to $(\mathcal{P} \to \mathcal{Q}) \wedge (\mathcal{Q} \to \mathcal{P})$, we see that $\mathcal{P} \equiv \mathcal{Q}$ if and only if both $\mathcal{Q} \Longrightarrow \mathcal{P}$ and $\mathcal{P} \Longrightarrow \mathcal{Q}$. Thus, we can show that two statements are logically equivalent if we can show that each of them can be logically deduced from the other. Also, we get a lot of rules about logical deduction for free—two rules of deduction for each logical equivalence we know. For example, since $\neg(p \wedge q) \equiv (\neg p \vee \neg q)$, we get that $\neg(p \wedge q) \Longrightarrow (\neg p \vee \neg q)$. For example, if we know "It is not both sunny and warm," then we can logically deduce "Either it's not sunny or it's not warm." (And vice versa.)

In general, arguments are more complicated that those we've considered so far. Here, for example, is an argument that has five premises:

$$\begin{array}{l} (p \wedge r) \rightarrow s \\ q \rightarrow p \\ t \rightarrow r \\ q \\ t \\ \hline \therefore\ s \end{array}$$

Is this argument valid? Of course, you could use a truth table to check whether the conjunction of the premises logically implies the conclusion. But with five propositional variables, the table would have 32 lines, and the size of the table grows quickly when more propositional variables are used. So, in general, truth tables are not practical.

Fortunately, there is another way to proceed, based on the fact that it is possible to chain several logical deductions together. That is, if $\mathcal{P} \Longrightarrow \mathcal{Q}$ and $\mathcal{Q} \Longrightarrow \mathcal{R}$, it follows that $\mathcal{P} \Longrightarrow \mathcal{R}$. This means we can demonstrate the validity of an argument by deducing the conclusion from the premises in a sequence of steps. These steps can be presented in the form of a proof:

Definition 1.10. A ***formal proof*** that an argument is valid consists of a sequence of propositions such that the last proposition in the sequence is the conclusion of the argument, and every proposition in the sequence is either a premise of the argument or follows by logical deduction from propositions that precede it in the list.

The existence of such a proof shows that the conclusion follows logically from the premises, and therefore that the argument is valid. Here is a formal proof that the argument given above is valid. The propositions in the proof are numbered, and each proposition has a justification.

1. $q \rightarrow p$	premise
2. q	premise
3. p	from 1 and 2 (*modus ponens*)
4. $t \rightarrow r$	premise
5. t	premise
6. r	from 4 and 5 (*modus ponens*)
7. $p \wedge r$	from 3 and 6
8. $(p \wedge r) \rightarrow s$	premise
9. s	from 7 and 8 (*modus ponens*)

Once a formal proof has been constructed, it is convincing. Unfortunately, it's not necessarily easy to come up with the proof. Usually, the best method is a combination of working forward ("Here's what I know, what can I deduce from that?") and working backwards ("Here's what I need to prove, what other things would imply that?"). For this proof, I might have thought: I want to prove s. I know that $p \wedge r$ implies s, so if I can prove $p \wedge r$, I'm OK. But to prove $p \wedge r$, it'll be enough to prove p and r separately....

Of course, not every argument is valid, so the question also arises, how can we show that an argument is invalid? Let's assume that the argument has been put into general form, with all the specific propositions replaced by propositional variables. The argument is valid if in all cases where all the premises are true, the conclusion is also true. The argument is invalid if there is even one case where all the premises are true and the conclusion is false. We can prove that an argument is invalid by finding an assignment of truth values to the propositional variables which makes all the premises true but makes the conclusion false. For example, consider an argument of the form:

$$\begin{array}{l} p \rightarrow q \\ q \rightarrow (p \wedge r) \\ r \\ \hline \therefore\ p \end{array}$$

In the case where p is false, q is false, and r is true, the three premises of this argument are all true, but the conclusion is false. This shows that the argument is invalid.

To apply all this to arguments stated in English, we have to introduce propositional variables to represent all the propositions in the argument. For example, consider:

> John will be at the party if Mary is there and Bill is not there. Mary will be at the party if it's on Friday or Saturday. If Bill is at the party, Tom will be there. Tom won't be at the party if it's on Friday. The party is on Friday. Therefore, John will be at the party.

Let j stand for "John will be at the party," m for "Mary will be there," b for "Bill will be there," t for "Tom will be there," f for "The party is on Friday," and s for "The party is on Saturday." Then this argument has the form

$$
\begin{array}{l}
(m \wedge \neg b) \rightarrow j \\
(f \vee s) \rightarrow m \\
b \rightarrow t \\
f \rightarrow \neg t \\
f \\
\hline
\therefore\ j
\end{array}
$$

This is a valid argument, as the following proof shows:

1.	$f \rightarrow \neg t$	premise
2.	f	premise
3.	$\neg t$	from 1 and 2 (*modus ponens*)
4.	$b \rightarrow t$	premise
5.	$\neg b$	from 4 and 3 (*modus tollens*)
6.	$f \vee s$	from 2
7.	$(f \vee s) \rightarrow m$	premise
8.	m	from 6 and 7 (*modus ponens*)
9.	$m \wedge \neg b$	from 8 and 5
10.	$(m \wedge \neg b) \rightarrow j$	premise
11.	j	from 10 and 9 (*modus ponens*)

So far in this section, we have been working mostly with propositional logic. But the definitions of valid argument and logical deduction apply to predicate logic as well. One of the most basic rules of deduction in predicate logic says that $(\forall x P(x)) \Longrightarrow P(a)$ for any entity a in the domain of discourse of the predicate P. That is, if a predicate is true of all entities, then it is true of any given particular entity. This rule can be combined with rules of deduction for propositional logic to give the following valid arguments

$$
\begin{array}{l}
\forall x(P(x) \rightarrow Q(x)) \\
P(a) \\
\hline
\therefore\ Q(a)
\end{array}
\qquad\qquad
\begin{array}{l}
\forall x(P(x) \rightarrow Q(x)) \\
\neg Q(a) \\
\hline
\therefore\ \neg P(a)
\end{array}
$$

These valid arguments go by the names of *modus ponens* and *modus tollens* for predicate logic. Note that from the premise $\forall x(P(x) \rightarrow Q(x))$ we can deduce $P(a) \rightarrow Q(a)$. From this and from the premise that $P(a)$, we can deduce $Q(a)$ by *modus ponens*. So the first argument above is valid. The second argument is similar, using *modus tollens*.

The most famous logical deduction of them all is an application of *modus ponens* for predicate logic:

$$\begin{array}{l} \text{All humans are mortal} \\ \text{Socrates is human} \\ \hline \therefore\ \text{Socrates is mortal} \end{array}$$

This has the form of *modus ponens* with $P(x)$ standing for "x is human," $Q(x)$ standing for "x is mortal," and a standing for the noted entity, Socrates.

There is a lot more to say about logical deduction and proof in predicate logic, and we'll spend the rest of this chapter on the subject.

Exercises

1. Verify the validity of *modus tollens* and the Law of Syllogism.

2. Each of the following is a valid rule of deduction. For each one, give an example of a valid argument in English that uses that rule.

$$\begin{array}{l} p \vee q \\ \neg p \\ \hline \therefore\ q \end{array} \qquad \begin{array}{l} p \\ q \\ \hline \therefore\ p \wedge q \end{array} \qquad \begin{array}{l} p \wedge q \\ \hline \therefore\ p \end{array} \qquad \begin{array}{l} p \\ \hline \therefore\ p \vee q \end{array}$$

3. There are two notorious invalid arguments that look deceptively like *modus ponens* and *modus tollens*:

$$\begin{array}{l} p \rightarrow q \\ q \\ \hline \therefore\ p \end{array} \qquad \begin{array}{l} p \rightarrow q \\ \neg p \\ \hline \therefore\ \neg q \end{array}$$

Show that each of these arguments is invalid. Give an English example that uses each of these arguments.

4. Decide whether each of the following arguments is valid. If it is valid, give a formal proof. If it is invalid, show that it is invalid by finding an appropriate assignment of truth values to propositional variables.

a) $$\begin{array}{l} p \rightarrow q \\ q \rightarrow s \\ s \\ \hline \therefore\ p \end{array}$$

b) $$\begin{array}{l} p \wedge q \\ q \rightarrow (r \vee s) \\ \neg r \\ \hline \therefore\ s \end{array}$$

c) $$\begin{array}{l} p \vee q \\ q \rightarrow (r \wedge s) \\ \neg p \\ \hline \therefore\ s \end{array}$$

d) $$\begin{array}{l} (\neg p) \rightarrow t \\ q \rightarrow s \\ r \rightarrow q \\ \neg(q \vee t) \\ \hline \therefore\ p \end{array}$$

e) $$\begin{array}{l} p \\ s \rightarrow r \\ q \vee r \\ q \rightarrow \neg p \\ \hline \therefore\ \neg s \end{array}$$

f) $$\begin{array}{l} q \rightarrow t \\ p \rightarrow (t \rightarrow s) \\ p \\ \hline \therefore\ q \rightarrow s \end{array}$$

5. For each of the following English arguments, express the argument in terms of propositional logic and determine whether the argument is valid or invalid.

a) If it is Sunday, it rains or snows. Today, it is Sunday and it's not raining. Therefore, it must be snowing.
b) If there are anchovies on the pizza, Jack won't eat it. If Jack doesn't eat pizza, he gets angry. Jack is angry. Therefore, there were anchovies on the pizza.
c) At 8:00, Jane studies in the library or works at home. It's 8:00 and Jane is not studying in the library. So she must be working at home.

1.6 Proof

Mathematics is unique in that it claims a certainty that is beyond all possible doubt or argument. A mathematical proof shows how some result follows by logic alone from a given set of assumptions, and once the result has been proven, it is as solid as the foundations of logic themselves. Of course, mathematics achieves this certainty by restricting itself to an artificial, mathematical world, and its application to the real world does not carry the same degree of certainty.

Within the world of mathematics, consequences follow from assumptions with the force of logic, and a proof is just a way of pointing out logical consequences. There is an old mathematical joke about this:

This mathematics professor walks into class one day and says "We'll start today with this result, which is obvious," and he writes it on the board. Then, he steps back and looks at the board for a while. He walks around the front of the room, stares into space and back at the board. This goes on till the end of class, and he walks out without saying anything else. The next class period, the professor walks into the room with a big smile, writes the same result on the board, turns to the class and says, "I was right. It is obvious."

For of course, the fact that mathematical results follow logically does not mean that they are obvious in any normal sense. Proofs are convincing once they are discovered, but finding them is often very difficult. They are written in a language and style that can seem obscure to the uninitiated. Often, a proof builds on a long series of definitions and previous results, and while each step along the way might be "obvious," the end result can be surprising and powerful. This is what makes the search for proofs worthwhile.

In the rest of this chapter, we'll look at some approaches and techniques that can be used for proving mathematical results, including two important proof techniques known as proof by contradiction and mathematical induction. Along the way, we'll encounter a few new definitions and notations. Hopefully, you will be left with a higher level of confidence for exploring the mathematical world on your own.

The mathematical world and the real world weren't always quite so separate. Until some time near the middle of the nineteenth century, the statements of mathematics were regarded as statements about the world. A proof was simply a convincing argument, rather than a chain forged of absolute logic. It was something closer to the original meaning of the word "proof", as a test or trial: To prove something was to test its truth by putting it to the trial of logical argument.

The first rumble of trouble came in the form of ***non-Euclidean geometry***. For two thousand years, the geometry of the Greek mathematician Euclid had been accepted, simply, as the geometry of the world. In the middle of the nineteenth century, it was discovered that there are other systems of geometry, which are at least as valid and self-consistent as Euclid's system. Mathematicians can work in any of these systems, but they cannot all claim to be working in the real world.

Near the end of the nineteenth century came another shock, in the form of cracks in the very foundation of mathematics. At that time, mathematician Gottlieb Frege was finishing a book on set theory that represented his life's work. In Frege's set theory, a set could be defined by any property. You could have, for example, the set consisting of all sets that contain three objects. As he was finishing his book, Frege received a letter from a young mathematician named Bertrand Russell which described what became known as ***Russell's Paradox***. Russell pointed out that the set of all sets—that is, the set that contains every entity that satisfies the property of being a set—cannot logically exist. We'll see Russell's reasoning in the following chapter. Frege could only include a postscript in his book stating that the basis of the work had been swept away.

Mathematicians responded to these problems by banishing appeals to facts about the real world from mathematical proof. Mathematics was to be its own world, built on its own secure foundation. The foundation would be a basic set of assumptions or "axioms" from which everything else would follow by logic. It would only be necessary to show that the axioms themselves were logically consistent and complete, and the world of mathematics would be secure. Unfortunately, even this was not to be. In the 1930s, Kurt Gödel showed that there is no consistent, finite set of axioms that completely describes even the corner of the mathematical world known as arithmetic. Gödel showed that given any finite, consistent set of axioms, there are true statements about arithmetic that do not follow logically from those axioms.

We are left with a mathematical world in which iron chains of logic still bind conclusions to assumptions. But the assumptions are no longer rooted in the real world. Nor is there any finite core of axioms to which the rest of the mathematical world can be chained. In this world, axioms are set up as signposts in a void, and then structures of logic are built around them. For example, instead of talking about *the* set theory that describes the real world, we have *a* set theory, based on a given set of axioms. That set theory is necessarily incomplete, and it might differ from other set theories which are based on other sets of axioms.

Understandably, mathematicians are very picky about getting their proofs right. It's how they construct their world. Students sometimes object that mathematicians are too picky about proving things that are "obvious." But the fact that something is obvious in the real world counts for very little in the constructed world of mathematics. Too many obvious things have turned out to be dead wrong. (For that matter, even things in the real world that seem "obviously" true are not necessarily true at all. For example, consider the quantity $f(n) = n^2 + n + 41$. When $n = 0$, $f(n) = 41$ which is prime; when $n = 1$, $f(n) = 43$ which is prime; when $n = 2$, $f(n) = 47$, which is prime. By the time you had calculated $f(3), f(4), \dots, f(10)$ and found that they were all prime, you might conclude that it is "obvious" that $f(n)$ is prime for all $n \geq 0$. But this is not in fact the case! (See exercises.) Similarly, those of you who are baseball fans might consider it "obvious" that if player A has a higher batting average against left-handers than player B, and player A has a higher batting average against right-handers than player B, then player A must have a higher batting average than player B. Again, this is not true!)

As we saw in Section 1.5, a formal proof consists of a sequence of statements where each statement is either an assumption or follows by a rule of logic from previous statements. The examples in that section all worked with unspecified generic propositions (p, q, etc). Let us now look at how one might use the same techniques to prove a specific proposition about the mathematical world. We will prove that for all integers n, if n is even then n^2 is even. (Definition: an integer n is *even* iff $n = 2k$ for some integer k. For example, 2 is even since $2 = 2 \cdot 1$; 66 is even since $66 = 2 \cdot 33$; 0 is even since $0 = 2 \cdot 0$.)

Proof. This is a proposition of the form $\forall n(P(n) \rightarrow Q(n))$ where $P(n)$ is "n is even" and $Q(n)$ is "n^2 is even." We need to show that $P(n) \rightarrow Q(n)$ is true for all values of n. In the language of Section 1.5, we need to show that for any n, $P(n)$ logically implies $Q(n)$; or, equivalently, that $Q(n)$ can be logically deduced from $P(n)$; or, equivalently, that

$$\begin{array}{l} n \text{ is even} \\ \hline \therefore\ n^2 \text{ is even} \end{array}$$

is a valid argument. Here is a formal proof that

$$\frac{n \text{ is even}}{\therefore\ n^2 \text{ is even}}$$

is in fact a valid argument for any value of n:

Let n be an arbitrary integer.

1. n is even	premise
2. if n is even, then $n = 2k$ for some integer k	definition of even
3. $n = 2k$ for some integer k	from 1, 2 (*modus ponens*)
4. if $n = 2k$ for some integer k, then $n^2 = 4k^2$ for that integer k	basic algebra
5. $n^2 = 4k^2$ for some integer k	from 3, 4 (*modus ponens*)
6. if $n^2 = 4k^2$ for some integer k, then $n^2 = 2(2k^2)$ for that k	basic algebra
7. $n^2 = 2(2k^2)$ for some integer k	from 5, 6 (*modus ponens*)
8. if $n^2 = 2(2k^2)$ for some integer k, then $n^2 = 2k'$ for some integer k'	basic fact about integers
9. $n^2 = 2k'$ for some integer k'	from 7, 8 (*modus ponens*)
10. if $n^2 = 2k'$ for some integer k', then n^2 is even	definition of even
11. n^2 is even	from 9, 10 (*modus ponens*)

(The "basic fact about integers" referred to above is that the product of integers is again an integer.) Since n could be replaced by any integer throughout this argument, we have proved the statement "if n is even then n^2 is even" is true for all integers n. (You might worry that the argument is only valid for even n; see the disclaimer about Keanu Reeves' acting ability on page 34, or remind yourself that $P(n) \to Q(n)$ is automatically true if $P(n)$ is false.) □

Mathematical proofs are rarely presented with this degree of detail and formality. A slightly less formal proof of our proposition might leave out the explicit implications and instances of *modus ponens* and appear as follows:

Proof. Let n be an arbitrary integer.

1. n is even	premise
2. $n = 2k$ for some integer k	definition of even
3. $n^2 = 4k^2$ for that integer k	basic algebra
4. $n^2 = 2(2k^2)$ for that k	basic algebra
5. $n^2 = 2k'$ for some integer k'	substituting $k' = 2k^2$
6. n^2 is even	definition of even

Since n was an arbitrary integer, the statement is true for all integers. □

A more typical proof would take the argument above and present it in prose rather than list form:

Proof. Let n be an arbitrary integer and assume n is even. Then $n = 2k$ for some integer k by the definition of even, and $n^2 = 4k^2 = 2(2k^2)$. Since the product of integers is an integer, we have $n^2 = 2k'$ for some integer k'. Therefore n^2 is even. Since n was an arbitrary integer, the statement is true for all integers. □

Typically, in a "formal" proof, it is this kind of (relatively) informal discussion that is given, with enough details to convince the reader that a complete, formal proof could be constructed. Of course, how many details the reader can be expected to fill in depends on the reader, and reading proofs is a skill that must be developed and practiced. Writing a proof is even more difficult. Every proof involves a creative act of discovery, in which a chain of logic that leads from assumptions to conclusion is discovered. It also involves a creative act of expression, in which that logic is presented in a clear and convincing way. There is no algorithm for producing correct, coherent proofs. There are, however, some general guidelines for discovering and writing proofs.

One of the most important pieces of advice to keep in mind is, "Use the definition." In the world of mathematics, terms mean exactly what they are defined to mean and nothing more. Definitions allow very complex ideas to be summarized as single terms. When you are trying to prove things about those terms, you generally need to "unwind" the definitions. In our example above, we used the definition of even to write $n = 2k$, and then we worked with that equation. When you are trying to prove something about equivalence relations in Chapter 2, you can be pretty sure that you will need to use the fact that equivalence relations, by definition, are symmetric, reflexive, and transitive. (And, of course, you'll need to know how the term "relation" is defined in the first place! You'll get nowhere if you work from the idea that "relations" are something like your aunt and uncle.)

More advice along the same line is to check whether you are using the assumptions of the theorem. An assumption that is made in a theorem is called an ***hypothesis***. The hypotheses of the theorem state conditions whose truth will guarantee the conclusion of the theorem. To prove the theorem means to assume that the hypotheses are true, and to show, under that assumption, that the conclusion must be true. It's likely (though not guaranteed) that you will need to use the hypotheses explicitly at some point in the proof, as we did in our example above.[12] Also, you should keep in mind that any result that has already been proved is available to be used in your proof.

A proof is a logical argument, based on the rules of logic. Since there are really not all that many basic rules of logic, the same patterns keep showing up over and over. Let's look at some of the patterns.

The most common pattern arises in the attempt to prove that something is true "for all" or "for every" or "for any" entity in a given category. In terms of logic, the statement you are trying to prove is of the form $\forall x\, P(x)$. In this case, the most likely way to begin the proof is by saying something like, "Let x be an arbitrary entity in the domain of discourse. We want to show that $P(x)$." In the rest of the proof, x refers to some unspecified but definite entity in the domain of discourse. Since x is arbitrary, proving $P(x)$ amounts to proving $\forall x\, P(x)$. You only have to be careful that you don't use any facts about x beyond what you have assumed. For example, in our proof above, we cannot make any assumptions about the integer n except that it is even; if we had made such assumptions, then the proof would have been incorrect, or at least incomplete.

Sometimes, you have to prove that an entity exists that satisfies certain stated properties. Such a proof is called an ***existence proof***. In this case, you are attempting to prove a statement of the form $\exists x\, P(x)$. The way to do this is to find an example, that is, to find a specific entity a for which $P(a)$ is true. One way to prove the statement "There is an even prime number" is to find a specific number that satisfies this description. The same statement could also be expressed "Not every prime number is odd." This statement has the form $\neg(\forall x\, P(x))$, which is equivalent to the statement $\exists x\, (\neg P(x))$. An example that proves the statement $\exists x\, (\neg P(x))$ also proves $\neg(\forall x\, P(x))$. Such an example is called a ***counterexample*** to the statement $\forall x\, P(x)$: A counterexample proves that the statement $\forall x\, P(x)$ is false. The number 2 is a counterexample to the statement "All prime numbers are odd." In

[12]Of course, if you set out to discover new theorems on your own, you aren't given the hypotheses and conclusion in advance, which makes things quite a bit harder—and more interesting.

fact, 2 is the only counterexample to this statement; many statements have multiple counterexamples.

Note that we have now discussed how to prove and disprove universally quantified statements, and how to prove existentially quantified statements. How do you *disprove* $\exists x\, P(x)$? Recall that $\neg\exists x\, P(x)$ is logically equivalent to $\forall x\, (\neg P(x))$, so to disprove $\exists x\, P(x)$ you need to prove $\forall x\, (\neg P(x))$.

Many statements, like that in our example above, have the logical form of an implication, $p \rightarrow q$. (More accurately, they are of the form "$\forall x\, (P(x) \rightarrow Q(x))$", but as discussed above the strategy for proving such a statement is to prove $P(x) \rightarrow Q(x)$ for an arbitrary element x of the domain of discourse.) The statement might be "For all natural numbers n, if n is even then n^2 is even," or "For all strings x, if x is in the language L then x is generated by the grammar G," or "For all elements s, if $s \in A$ then $s \in B$." Sometimes the implication is implicit rather than explicit: for example, "The sum of two rationals is rational" is really short for "For any numbers x and y, if x and y are rational then $x + y$ is rational." A proof of such a statement often begins something like this: "Assume that p. We want to show that q." In the rest of the proof, p is treated as an assumption that is known to be true. As discussed above, the logical reasoning behind this is that you are essentially proving that

$$\begin{array}{l} p \\ \hline \therefore\ q \end{array}$$

is a valid argument. Another way of thinking about it is to remember that $p \rightarrow q$ is automatically true in the case where p is false, so there is no need to handle that case explicitly. In the remaining case, when p is true, we can show that $p \rightarrow q$ is true by showing that the truth of q follows from the truth of p.

A statement of the form $p \wedge q$ can be proven by proving p and q separately. A statement of the form $p \vee q$ can be proved by proving the logically equivalent statement $(\neg p) \rightarrow q$: to prove $p \vee q$, you can assume that p is false and prove, under that assumption, that q is true. For example, the statement "Every integer is either even or odd" is equivalent to the statement "Every integer that is not even is odd."

Since $p \leftrightarrow q$ is equivalent to $(p \rightarrow q) \wedge (q \rightarrow p)$, a statement of the form $p \leftrightarrow q$ is often proved by giving two proofs, one of $p \rightarrow q$ and one of $q \rightarrow p$. In English, $p \leftrightarrow q$ can be stated in several forms such as "p if and only if q", "if p then q and conversely," and "p is necessary and sufficient for q." The phrase "if and only if" is so common in mathematics that it is often abbreviated ***iff***.

You should also keep in mind that you can prove $p \to q$ by displaying a chain of valid implications $p \to r \to s \to \cdots \to q$. Similarly, $p \leftrightarrow q$ can be proved with a chain of valid biconditionals $p \leftrightarrow r \leftrightarrow s \leftrightarrow \cdots \leftrightarrow q$.

We'll turn to a few examples, but first here is some terminology that we will use throughout our sample proofs:

- The ***natural numbers*** (denoted $\mathbb{N}$) are the numbers $0, 1, 2, \ldots$. Note that the sum and product of natural numbers are natural numbers.
- The ***integers*** (denoted $\mathbb{Z}$) are the numbers $0, -1, 1, -2, 2, -3, 3, \ldots$. Note that the sum, product, and difference of integers are integers.
- The ***rational numbers*** (denoted $\mathbb{Q}$) are all numbers that can be written in the form $\frac{m}{n}$ where m and n are integers and $n \neq 0$. So $\frac{1}{3}$ and $\frac{-65}{7}$ are rationals; so, less obviously, are 6 and $\frac{\sqrt{27}}{\sqrt{12}}$ since $6 = \frac{6}{1}$ (or, for that matter, $6 = \frac{-12}{-2}$), and $\frac{\sqrt{27}}{\sqrt{12}} = \sqrt{\frac{27}{12}} = \sqrt{\frac{9}{4}} = \frac{3}{2}$. Note the restriction that the number in the denominator cannot be 0: $\frac{3}{0}$ is not a number at all, rational or otherwise; it is an undefined quantity. Note also that the sum, product, difference, and quotient of rational numbers are rational numbers (provided you don't attempt to divide by 0.)
- The ***real numbers*** (denoted $\mathbb{R}$) are numbers that can be written in decimal form, possibly with an infinite number of digits after the decimal point. Note that the sum, product, difference, and quotient of real numbers are real numbers (provided you don't attempt to divide by 0.)
- The ***irrational numbers*** are real numbers that are not rational, i.e. that cannot be written as a ratio of integers. Such numbers include $\sqrt{3}$ (which we will prove is not rational) and π (if anyone ever told you that $\pi = \frac{22}{7}$, they lied—$\frac{22}{7}$ is only an *approximation* of the value of π).
- An integer n is ***divisible by*** m iff $n = mk$ for some integer k. (This can also be expressed by saying that m evenly divides n.) So for example, n is divisible by 2 iff $n = 2k$ for some integer k; n is divisible by 3 iff $n = 3k$ for some integer k, and so on. Note that if n is *not* divisible by 2, then n must be 1 more than a multiple of 2 so $n = 2k+1$ for some integer k. Similarly, if n is not divisible by 3 then n must be 1 or 2 more than a multiple of 3, so $n = 2k + 1$ or $n = 2k + 2$ for some integer k.

- An integer is ***even*** iff it is divisible by 2 and ***odd*** iff it is not.
- An integer $n > 1$ is ***prime*** if it is divisible by exactly two positive integers, namely 1 and itself. Note that a number must be greater than 1 to even have a chance of being termed "prime". In particular, neither 0 nor 1 is prime.

Let's look now at another example: prove that the sum of any two rational numbers is rational.

Proof. We start by assuming that x and y are arbitrary rational numbers. Here's a formal proof that the inference rule

$$\begin{array}{l} x \text{ is rational} \\ \underline{y \text{ is rational}} \\ \therefore\ x + y \text{ is rational} \end{array}$$

is a valid rule of inference:

1.	x is rational	premise
2.	if x is rational, then $x = \frac{a}{b}$ for some integers a and $b \neq 0$	definition of rationals
3.	$x = \frac{a}{b}$ for some integers a and $b \neq 0$	from 1,2 (*modus ponens*)
4.	y is rational	premise
5.	if y is rational, then $y = \frac{c}{d}$ for some integers c and $d \neq 0$	definition of rational
6.	$y = \frac{c}{d}$ for some c and $d \neq 0$	from 4,5 (*modus ponens*)
7.	$x = \frac{a}{b}$ for some a and $b \neq 0$ and $y = \frac{c}{d}$ for some c and $d \neq 0$	from 3,6
8.	if $x = \frac{a}{b}$ for some a and $b \neq 0$ and $y = \frac{c}{d}$ for c and $d \neq 0$ then $x + y = \frac{ad+bc}{bd}$ where a, b, c, d are integers and $b, d \neq 0$	basic algebra
9.	$x + y = \frac{ad+bc}{bd}$ for some a, b, c, d where $b, d \neq 0$	from 7,8 (*modus ponens*)
10.	if $x + y = \frac{ad+bc}{bd}$ for some a, b, c, d where $b, d \neq 0$ then $x + y = \frac{m}{n}$ where m, n are integers and $n \neq 0$	properties of integers
11.	$x + y = \frac{m}{n}$ where m and n are integers and $n \neq 0$	from 9,10 (*modus ponens*)
12.	if $x + y = \frac{m}{n}$ where m and n are integers and $n \neq 0$ then $x + y$ is rational	definition of rational
13.	$x + y$ is rational	from 11,12 (*modus ponens*)

So the rule of inference given above is valid. Since x and y are arbitrary rationals, we have proved that the rule is valid for all rationals, and hence the sum of any two rationals is rational. □

Again, a more informal presentation would look like:

Proof. Let x and y be arbitrary rational numbers. By the definition of rational, there are integers $a, b \neq 0, c, d \neq 0$ such that $x = \frac{a}{b}$ and $y = \frac{c}{d}$. Then $x + y = \frac{ad+bc}{bd}$; we know $ad + bc$ and bd are integers since the sum and product of integers are integers, and we also know $bd \neq 0$ since neither b nor d is 0. So we have written $x + y$ as the ratio of two integers, the denominator being non-zero. Therefore, by the definition of rational numbers, $x + y$ is rational. Since x and y were arbitrary rationals, the sum of any two rationals is rational. □

And one more example: we will prove that any 4-digit number $d_1d_2d_3d_4$ is divisible by 3 iff the sum of the four digits is divisible by 3.

Proof. This statement is of the form $p \leftrightarrow q$; recall that $p \leftrightarrow q$ is logically equivalent to $(p \rightarrow q) \wedge (q \rightarrow p)$. So we need to prove for any 4-digit number $d_1d_2d_3d_4$ that (1) if $d_1d_2d_3d_4$ is divisible by 3 then $d_1+d_2+d_3+d_4$ is divisible by 3, and (2) if $d_1+d_2+d_3+d_4$ is divisible by 3 then $d_1d_2d_3d_4$ is divisible by 3. So let $d_1d_2d_3d_4$ be an arbitrary 4-digit number.

(1) Assume $d_1d_2d_3d_4$ is divisible by 3, i.e. $d_1d_2d_3d_4 = 3k$ for some integer k. The number $d_1d_2d_3d_4$ is actually $d_1 \times 1000 + d_2 \times 100 + d_3 \times 10 + d_4$, so we have the equation

$$d_1 \times 1000 + d_2 \times 100 + d_3 \times 10 + d_4 = 3k.$$

Since $1000 = 999 + 1$, $100 = 99 + 1$, and $10 = 9 + 1$, this equation can be rewritten

$$999d_1 + d_1 + 99d_2 + d_2 + 9d_3 + d_3 + d_4 = 3k.$$

Rearranging gives

$$\begin{aligned} d_1 + d_2 + d_3 + d_4 &= 3k - 999d_1 - 99d_2 - 9d_3 \\ &= 3k - 3(333d_1) - 3(33d_2) - 3(3d_3). \end{aligned}$$

We can now factor a 3 from the right side to get

$$d_1 + d_2 + d_3 + d_4 = 3(k - 333d_1 - 33d_2 - d_3).$$

Since $(k - 333d_1 - 33d_2 - d_3)$ is an integer, we have shown that $d_1+d_2+d_3+d_4$ is divisible by 3.

(2) Assume $d_1 + d_2 + d_3 + d_4$ is divisible by 3. Consider the number $d_1d_2d_3d_4$. As remarked above,

$$d_1d_2d_3d_4 = d_1 \times 1000 + d_2 \times 100 + d_3 \times 10 + d_4$$

so

$$\begin{aligned} d_1d_2d_3d_4 &= 999d_1 + d_1 + 99d_2 + d_2 + 9d_3 + d_3 + d_4 \\ &= 999d_1 + 99d_2 + 9d_3 + (d_1 + d_2 + d_3 + d_4). \end{aligned}$$

We assumed that $d_1 + d_2 + d_3 + d_4 = 3k$ for some integer k, so we can substitute into the last equation to get

$$d_1d_2d_3d_4 = 999d_1 + 99d_2 + 9d_3 + 3k = 3(333d_1 + 33d_2 + 3d_3 + k).$$

Since the quantity in parentheses is an integer, we have proved that $d_1d_2d_3d_4$ is divisible by 3.

In (1) and (2) above, the number $d_1d_2d_3d_4$ was an arbitrary 4-digit integer, so we have proved that for all 4-digit integers, $d_1d_2d_3d_4$ is divisible by 3 iff the sum of the four digits is divisible by 3. $\square$

Now suppose we wanted to prove the statement "For all integers n, n^2 is even if and only if n is even." We have already proved half of this statement ("For all integers n, if n is even then n^2 is even"), so all we need to do is prove the statement "For all integers n, if n^2 is even then n is even" and we'll be done. Unfortunately, this is not as straightforward as it seems: suppose we started in our standard manner and let n be an arbitrary integer and assumed that $n^2 = 2k$ for some integer k. Then we'd be stuck! Taking the square root of both sides would give us n on the left but would leave a $\sqrt{2k}$ on the right. This quantity is not of the form $2k'$ for any integer k'; multiplying it by $\frac{\sqrt{2}}{\sqrt{2}}$ would give $2\frac{\sqrt{k}}{\sqrt{2}}$ but there is no way for us to prove that $\frac{\sqrt{k}}{\sqrt{2}}$ is an integer. So we've hit a dead end. What do we do now?

The answer is that we need a different proof technique. The proofs we have written so far are what are called ***direct proofs***: to prove $p \to q$ you assume p is true and prove that the truth of q follows. Sometimes, when a direct proof of $p \to q$ fails, an ***indirect proof*** will work. Recall that the *contrapositive* of the implication $p \to q$ is the implication $\neg q \to \neg p$, and that this proposition is logically equivalent to $p \to q$. An indirect proof of $p \to q$, then, is a direct proof of the contrapositive $\neg q \to \neg p$. In our current example, instead of proving "if n^2 is even then n is even" directly, we can prove its contrapositive "if n is not even (i.e. n is odd) then n^2 is not even (i.e. n^2 is odd.)" The proof of this contrapositive is a routine direct argument which we leave to the exercises.

Exercises

1. Find a natural number n for which $n^2 + n + 41$ is not prime.

2. Show that the propositions $p \vee q$ and $(\neg p) \to q$ are logically equivalent.

3. Show that the proposition $(p \vee q) \to r$ is equivalent to $(p \to r) \wedge (q \to r)$.

4. Determine whether each of the following statements is true. If it true, prove it. If it is false, give a counterexample.

a) Every prime number is odd.

b) Every prime number greater than 2 is odd.

c) If x and y are integers with $x < y$, then there is an integer z such that $x < z < y$.
d) If x and y are real numbers with $x < y$, then there is a real number z such that $x < z < y$.

5. Suppose that r, s, and t are integers, such that r evenly divides s and s evenly divides t. Prove that r evenly divides t.
6. Prove that for all integers n, if n is odd then n^2 is odd.
7. Prove that an integer n is divisible by 3 iff n^2 is divisible by 3. (Hint: give an indirect proof of "if n^2 is divisible by 3 then n is divisible by 3.")
8. Prove or disprove each of the following statements.
   a) The product of two even integers is even.
   b) The product of two integers is even only if both integers are even.
   c) The product of two rational numbers is rational.
   d) The product of two irrational numbers is irrational.
   e) For all integers n, if n is divisible by 4 then n^2 is divisible by 4.
   f) For all integers n, if n^2 is divisible by 4 then n is divisible by 4.

1.7 Proof by Contradiction

Suppose that we start with some set of assumptions and apply rules of logic to derive a sequence of statements that can be proved from those assumptions, and suppose that we derive a statement that we know to be false. When the laws of logic are applied to true statements, the statements that are derived will also be true. If we derive a false statement by applying rules of logic to a set of assumptions, then at least one of the assumptions must be false. This observation leads to a powerful proof technique, which is known as ***proof by contradiction***.

Suppose that you want to prove some proposition, p. To apply proof by contradiction, assume that $\neg p$ is true, and apply the rules of logic to derive conclusions based on this assumption. If it is possible to derive a statement that is known to be false, it follows that the assumption, $\neg p$, must be false. (Of course, if the derivation is based on several assumptions, then you only know that at least *one* of the assumptions must be false.) The fact that $\neg p$ is false proves that p is true. Essentially, you are arguing that p must be true, because if it weren't, then some statement that is known to be false could be proved to be true. Generally, the false statement that is derived in a proof by contradiction is of the form $q \wedge \neg q$. This statement is a contradiction in the sense that it is false no matter what the value of q. Note that deriving the contradiction $q \wedge \neg q$ is the same as showing that the two statements, q and $\neg q$, both follow from the assumption that $\neg p$.

As a first example of proof by contradiction, consider the following theorem:

Theorem 1.4. *The number* $\sqrt{3}$ *is irrational.*

Proof. Assume for the sake of contradiction that $\sqrt{3}$ is rational. Then $\sqrt{3}$ can be written as the ratio of two integers, $\sqrt{3} = \frac{m'}{n'}$ for some integers m' and n'. Furthermore, the fraction $\frac{m'}{n'}$ can be reduced to lowest terms by canceling all common factors of m' and n'. So $\sqrt{3} = \frac{m}{n}$ for some integers m and n which have no common factors. Squaring both sides of this equation gives $3 = \frac{m^2}{n^2}$ and re-arranging gives $3n^2 = m^2$. From this equation we see that m^2 is divisible by 3; you proved in the previous section (Exercise 6) that m^2 is divisible by 3 iff m is divisible by 3. Therefore m is divisible by 3 and we can write $m = 3k$ for some integer k. Substituting $m = 3k$ into the last equation above gives $3n^2 = (3k)^2$ or $3n^2 = 9k^2$, which in turn becomes $n^2 = 3k^2$. From this we see that n^2 is divisible by 3, and again we know that this implies that n is divisible by 3. But now we have (i) m and n have no common factors, and (ii) m and n have a common factor, namely 3. It is impossible for both these things to be true, yet our argument has been logically correct. Therefore our original assumption, namely that $\sqrt{3}$ is rational, must be incorrect. Therefore $\sqrt{3}$ must be irrational. □

One of the oldest mathematical proofs, which goes all the way back to Euclid, is a proof by contradiction. Recall that a prime number is an integer n, greater than 1, such that the only positive integers that evenly divide n are 1 and n. We will show that there are infinitely many primes. Before we get to the theorem, we need a lemma. (A ***lemma*** is a theorem that is introduced only because it is needed in the proof of another theorem. Lemmas help to organize the proof of a major theorem into manageable chunks.)

Lemma 1.5. *If* N *is an integer and* $N > 1$*, then there is a prime number which evenly divides* N.

Proof. Let D be the smallest integer which is greater than 1 and which evenly divides N. (D exists since there is at least one number, namely N itself, which is greater than 1 and which evenly divides N. We use the fact that any non-empty subset of $\mathbb{N}$ has a smallest member.) I claim that D is prime, so that D is a prime number that evenly divides N.

Suppose that D is not prime. We show that this assumption leads to a contradiction. Since D is not prime, then, by definition, there is a number k between 2 and $D - 1$, inclusive, such that k evenly divides D. But since D evenly divides N, we also have that k evenly divides N (by exercise 5 in the previous section). That is, k is an integer greater than one which evenly divides N. But since k is less than D, this contradicts the fact that D is the *smallest* such number. This contradiction proves that D is a prime number. □

Theorem 1.6. *There are infinitely many prime numbers.*

Proof. Suppose that there are only finitely many prime numbers. We will show that this assumption leads to a contradiction.

Let p_1, p_2, ..., p_n be a complete list of all prime numbers (which exists under the assumption that there are only finitely many prime numbers). Consider the number N obtained by multiplying all the prime numbers together and adding one. That is,

$$N = (p_1 \cdot p_2 \cdot p_3 \cdots p_n) + 1.$$

Now, since N is larger than any of the prime numbers p_i, and since p_1, p_2, ..., p_n is a *complete* list of prime numbers, we know that N cannot be prime. By the lemma, there is a prime number p which evenly divides N. Now, p must be one of the numbers p_1, p_2, ..., p_n. But in fact, none of these numbers evenly divides N, since dividing N by any p_i leaves a remainder of 1. This contradiction proves that the assumption that there are only finitely many primes is false. □

This proof demonstrates the power of proof by contradiction. The fact that is proved here is not at all obvious, and yet it can be proved in just a few paragraphs.

Exercises

1. Suppose that a_1, a_2, ..., a_{10} are real numbers, and suppose that $a_1 + a_2 + \cdots + a_{10} > 100$. Use a proof by contradiction to conclude that at least one of the numbers a_i must be greater than 10.

2. Prove that each of the following statements is true. In each case, use a proof by contradiction. Remember that the negation of $p \to q$ is $p \wedge \neg q$.

a) Let n be an integer. If n^2 is an even integer, then n is an even integer.

b) $\sqrt{2}$ is irrational.

c) If r is a rational number and x is an irrational number, then $r + x$ is an irrational number. (That is, the sum of a rational number and an irrational number is irrational.)

d) If r is a non-zero rational number and x is an irrational number, then rx is an irrational number.

e) If r and $r + x$ are both rational, then x is rational.

3. The ***pigeonhole principle*** is the following obvious observation: If you have n pigeons in k pigeonholes and if $n > k$, then there is at least one pigeonhole that contains more than one pigeon. Even though this observation seems obvious, it's a good idea to prove it. Prove the pigeonhole principle using a proof by contradiction.

1.8 Mathematical Induction

The structure of the natural numbers—0, 1, 2, 3, and on to infinity—makes possible a powerful proof technique known as ***induction*** or ***mathematical induction***. The idea behind induction is simple. Let P be a one-place predicate whose domain of discourse includes the natural numbers. Suppose that we can prove that $P(0)$ is true. Suppose that we can also prove the statements $P(0) \rightarrow P(1)$, $P(1) \rightarrow P(2)$, $P(2) \rightarrow P(3)$, and so on. The principal of mathematical induction is the observation that we can then conclude that $P(n)$ is true for *all* natural numbers n. This should be clear. Since $P(0)$ and $P(0) \rightarrow P(1)$ are true, we can apply the rule of *modus ponens* to conclude that $P(1)$ is true. Then, since $P(1)$ and $P(1) \rightarrow P(2)$ are true, we can conclude by *modus ponens* that $P(2)$ is true. From $P(2)$ and $P(2) \rightarrow P(3)$, we conclude that $P(3)$ is true. For any given n in the set $\mathbb{N}$, we can continue this chain of deduction for n steps to prove that $P(n)$ is true.

When applying induction, we don't actually prove each of the implications $P(0) \rightarrow P(1)$, $P(1) \rightarrow P(2)$, and so on, individually. That would require an infinite amount of work. The whole point of induction is to avoid any infinitely long process. Instead, we prove $\forall k\,(P(k) \rightarrow P(k+1))$ (where the domain of discourse for the predicate P is $\mathbb{N}$). The statement $\forall k\,(P(k) \rightarrow P(k+1))$ summarizes all the infinitely many implications in a single statement. Stated formally, the principle of mathematical induction says that if we can prove the statement $P(0) \wedge \big(\forall k\,(P(k) \rightarrow P(k+1))\big)$, then we can deduce that $\forall n\, P(n)$ (again, with $\mathbb{N}$ as the domain of discourse).

It should be intuitively clear that the principle of induction is valid. It follows from the fact that the list 0, 1, 2, 3, ..., if extended long enough, will eventually include any given natural number. If we start from $P(0)$ and take enough steps of the form $P(k) \rightarrow P(k+1)$, we can get $P(n)$ for any given natural number n. However, whenever we deal with infinity, we are courting the possibility of paradox. We will prove the principle of induction rigorously in the next chapter (see Theorem 2.3), but for now we just state it as a theorem:

Theorem 1.7. *Let P be a one-place predicate whose domain of discourse* *includes the natural numbers. Suppose that* $P(0) \wedge \big(\forall k \in \mathbb{N}\,(P(k) \rightarrow P(k+1))\big)$. *Then* $P(n)$ *is true for all natural numbers* n. *(That is, the statement* $\forall n\, P(n)$ *is true, where the domain of discourse for* P *is the set of natural numbers.)*

Mathematical induction can be applied in many situations: you can prove things about strings of characters by doing induction on the length of the string, things about graphs by doing induction on the number of nodes in the graph, things about grammars by doing induction on the number of productions in the grammar, and so on. We'll be looking at applications of induction for the rest of this chapter, and throughout the remainder of the text. Although proofs by induction can be very different from one another, they all follow just a few basic structures. A proof based on the preceding theorem always has two parts. First, $P(0)$ is proved. This is called the ***base case*** of the induction. Then the statement $\forall k\,(P(k) \rightarrow P(k+1))$ is proved. This statement can be proved by letting k be an arbitrary element of $\mathbb{N}$ and proving $P(k) \rightarrow P(k+1)$. This in turn can be proved by assuming that $P(k)$ is true and proving that the truth of $P(k+1)$ follows from that assumption. This case is called the ***inductive case***, and $P(k)$ is called the ***inductive hypothesis*** or the ***induction hypothesis***. Note that the base case is just as important as the inductive case. By itself, the truth of the statement $\forall k\,(P(k) \rightarrow P(k+1))$ says nothing at all about the truth of any of the individual statements $P(n)$. The chain of implications $P(0) \rightarrow P(1)$, $P(1) \rightarrow P(2)$, …, $P(n-1) \rightarrow P(n)$ says nothing about $P(n)$ unless the chain is anchored at the other end by the truth of $P(0)$. Let's look at a few examples.

Theorem 1.8. *The number* $2^{2n}-1$ *is divisible by 3 for all natural numbers* n.

Proof. Here, $P(n)$ is the statement that $2^{2n}-1$ is divisible by 3.

Base case: When $n = 0$, $2^{2n} - 1 = 2^0 - 1 = 1 - 1 = 0$ and 0 is divisible by 3 (since $0 = 3 \cdot 0$.) Therefore the statement holds when $n = 0$.

Inductive case: We want to show that if the statement is true for $n = k$ (where k is an arbitrary natural number), then it is true for $n = k+1$ also. That is, we must prove the implication $P(k) \rightarrow P(k+1)$. So we assume $P(k)$, that is, we assume that 2^{2k} is divisible by 3. This means that $2^{2k} - 1 = 3m$ for some integer m. We want to prove $P(k+1)$, that is, that

$2^{2(k+1)} - 1$ is also divisible by 3:

$$\begin{aligned}
2^{2(k+1)} - 1 &= 2^{2k+2} - 1 \\
&= 2^{2k} \cdot 2^2 - 1 && \text{properties of exponents} \\
&= 4 \cdot 2^{2k} - 1 \\
&= 4 \cdot 2^{2k} - 4 + 4 - 1 \\
&= 4(2^{2k} - 1) + 3 && \text{algebra} \\
&= 4(3m) + 3 && \text{the inductive hypothesis} \\
&= 3(4m + 1) && \text{algebra}
\end{aligned}$$

and from the last line we see that 2^{2k+1} is in fact divisible by 3. (The third step—subtracting and adding 4—was done to enable us to use our inductive hypothesis.)

Altogether, we have proved that $P(0)$ holds and that, for all k, $P(k) \rightarrow P(k+1)$ is true. Therefore, by the principle of induction, $P(n)$ is true for all n in $\mathbb{N}$, i.e. $2^{2n} - 1$ is divisible by 3 for all n in $\mathbb{N}$. □

The principal of mathematical induction gives a method for proving $P(n)$ for all n in the set $\mathbb{N}$. It should be clear that if M is any natural number, a similar method can be used to show that $P(n)$ is true for all natural numbers n that satisfy $n \geq M$. Just start the induction with a base case of $n = M$ instead of with a base case of $n = 0$. I leave the proof of this extension of the principle of induction as an exercise. We can use the extended principle of induction to prove a result that was first mentioned in Section 1.1.

Theorem 1.9. *Suppose that a compound proposition contains exactly n propositional variables, where $n \geq 1$. Then there are exactly 2^n different ways of assigning truth values to the n variables.*

Proof. Let $P(n)$ be the statement "There are exactly 2^n different ways of assigning truth values to n propositional variables." We will use induction to prove the $P(n)$ is true for all $n \geq 1$.

Base case: First, we prove the statement $P(1)$. If there is exactly one variable, then there are exactly two ways of assigning a truth value to that variable. Namely, the variable can be either *true* or *false.* Since $2 = 2^1$, $P(1)$ is true.

Inductive case: Suppose that $P(k)$ is already known to be true. We want to prove that, under this assumption, $P(k+1)$ is also true. Suppose that $p_1, p_2, \ldots, p_{k+1}$ are $k+1$ propositional variables. Since we are assuming that $P(k)$ is true, we know that there are 2^k ways of assigning truth values to p_1, p_2, $\ldots$, p_k. But each assignment of truth values to p_1, p_2, $\ldots$, p_k can be extended to the complete list p_1, p_2, $\ldots$, p_k, p_{k+1} in two ways. Namely, p_{k+1} can be assigned the value *true* or the value *false*. It follows that there are $2 \cdot 2^k$ ways of assigning truth values to p_1, p_2, $\ldots$, p_{k+1}. Since $2 \cdot 2^k = 2^{k+1}$, this finishes the proof. □

The sum of an arbitrary number of terms is written using the symbol $\sum$. (This symbol is the Greek letter sigma, which is equivalent to the Latin letter S and stands for "sum.") Thus, we have

$$\sum_{i=1}^{5} i^2 = 1^2 + 2^2 + 3^2 + 4^2 + 5^2$$

$$\sum_{k=3}^{7} a_k = a_3 + a_4 + a_5 + a_6 + a_7$$

$$\sum_{n=0}^{N} \frac{1}{n+1} = \frac{1}{0+1} + \frac{1}{1+1} + \frac{1}{2+1} + \cdots + \frac{1}{N+1}$$

This notation for a sum, using the $\sum$ operator, is called ***summation notation***. A similar notation for products uses the symbol $\prod$. (This is the Greek letter pi, which is equivalent to the Latin letter P and stands for "product.") For example,

$$\prod_{k=2}^{5} (3k+2) = (3 \cdot 2 + 2)(3 \cdot 3 + 2)(3 \cdot 4 + 2)(3 \cdot 5 + 2)$$

$$\prod_{i=1}^{n} \frac{1}{i} = \frac{1}{1} \cdot \frac{1}{2} \cdots \frac{1}{n}.$$

Induction can be used to prove many formulas that use these notations. Here are two examples:

Theorem 1.10. $\displaystyle\sum_{i=1}^{n} i = \frac{n(n+1)}{2}$ *for any integer n greater than zero.*

Proof. Let $P(n)$ be the statement $\displaystyle\sum_{i=1}^{n} i = \frac{n(n+1)}{2}$. We use induction to show that $P(n)$ is true for all $n \geq 1$.

Base case: Consider the case $n = 1$. $P(1)$ is the statement that $\sum_{i=1}^{1} i = \frac{1(1+1)}{2}$. Since $\sum_{i=1}^{1} i = 1$ and $\frac{1(1+1)}{2} = 1$, $P(1)$ is true.

Inductive case: Let $k > 1$ be arbitrary, and assume that $P(k)$ is true. We want to show that $P(k+1)$ is true. $P(k+1)$ is the statement $\sum_{i=1}^{k+1} i = \frac{(k+1)(k+2)}{2}$. But

$$\begin{aligned}
\sum_{i=1}^{k+1} i &= \left(\sum_{i=1}^{k} i\right) + (k+1) \\
&= \frac{k(k+1)}{2} + (k+1) && \text{(inductive hypothesis)} \\
&= \frac{k(k+1)}{2} + \frac{2(k+1)}{2} \\
&= \frac{k(k+1) + 2(k+1)}{2} \\
&= \frac{(k+2)(k+1)}{2} \\
&= \frac{(k+1)(k+2)}{2}
\end{aligned}$$

which is what we wanted to show. This computation completes the induction. □

Theorem 1.11. $\sum_{i=1}^{n} i2^{i-1} = (n-1) \cdot 2^n + 1$ *for any natural number* $n > 0$.

Proof. Let $P(n)$ be the statement $\sum_{i=1}^{n} i2^{i-1} = (n-1) \cdot 2^n + 1$. We use induction to show that $P(n)$ is true for all $n > 0$

Base case: Consider the case $n = 1$. $P(1)$ is the statement that $\sum_{i=1}^{1} i2^{i-1} = (1-1) \cdot 2^1 + 1$. Since each side of this equation is equal to one, this is true.

Inductive case: Let $k > 1$ be arbitrary, and assume that $P(k)$ is true.

We want to show that $P(k+1)$ is true. $P(k+1)$ is the statement $\sum_{i=1}^{k+1} i2^{i-1} = ((k+1)-1)\cdot 2^{k+1}+1$. But, we can compute that

$$\begin{aligned}
\sum_{i=1}^{k+1} i2^{i-1} &= \left(\sum_{i=1}^{k} i2^{i-1}\right) + (k+1)2^{(k+1)-1} \\
&= \left((k-1)\cdot 2^k + 1\right) + (k+1)2^k \qquad \text{(inductive hypothesis)} \\
&= \left((k-1)+(k+1)\right)2^k + 1 \\
&= (k\cdot 2)\cdot 2^k + 1 \\
&= k2^{k+1} + 1
\end{aligned}$$

which is what we wanted to show. This completes the induction. □

For example, these theorems show that $\sum_{i=1}^{100} i = 1+2+3+4+\cdots+100 = \frac{100(100+1)}{2} = 5050$ and that $1\cdot 2^0+2\cdot 2^1+3\cdot 2^2+4\cdot 2^3+5\cdot 2^4 = (5-1)2^5+1 = 129$, as well as infinitely many other such sums.

There is a second form of the principle of mathematical induction which is useful in some cases. To apply the first form of induction, we assume $P(k)$ for an arbitrary natural number k and show that $P(k+1)$ follows from that assumption. In the second form of induction, the assumption is that $P(x)$ holds for all x between 0 and k inclusive, and we show that $P(k+1)$ follows from this. This gives us a lot more to work with when deducing $P(k+1)$. We will need this second form of induction in the next two sections. A proof will be given in the next chapter.

Theorem 1.12. *Let P be a one-place predicate whose domain of discourse includes the natural numbers. Suppose that $P(0)$ is true and that*

$$(P(0) \wedge P(1) \wedge \cdots \wedge P(k)) \rightarrow P(k+1)$$

is true for each natural number $k \geq 0$. Then $P(n)$ is true for every natural number n.

For example, we can use this theorem to prove that every integer greater than one can be written as a product of prime numbers (where a number that is itself prime is considered to be a product of one prime number). The proof illustrates an important point about applications of this theorem:

When proving $P(k+1)$, you don't necessarily have to *use* the assumptions that $P(0)$, $P(1)$, ..., and $P(k)$ are true. If $P(k+1)$ is proved by *any* means—possibly including the assumptions—then the statement $(P(0) \wedge P(1) \wedge \cdots \wedge P(k)) \rightarrow P(k+1)$ has been shown to be true. It follows from this observation that several numbers, not just zero, can be "base cases" in the sense that $P(x+1)$ can be proved independently of $P(0)$ through $P(x)$. In this sense, 0, 1, and every prime number are base cases in the following theorem.

Theorem 1.13. *Every natural number greater than one can be written as a product of prime numbers.*

Proof. Let $P(n)$ be the statement "if $n > 1$, then n can be written as a product of prime numbers." We will prove that $P(n)$ is true for all n by applying the second form of the principle of induction.

Note that $P(0)$ and $P(1)$ are both automatically true, since $n = 0$ and $n = 1$ do not satisfy the condition that $n > 1$, and $P(2)$ is true since 2 is the product of the single prime number 2. Suppose that k is an arbitrary natural number with $k > 1$, and suppose that $P(0)$, $P(1)$, ..., $P(k)$ are already known to be true; we want to show that $P(k+1)$ is true. In the case where $k+1$ is a prime number, then $k+1$ is a product of one prime number, so $P(k+1)$ is true.

Consider the case where $k+1$ is not prime. Then, according to the definition of prime number, it is possible to write $k+1 = ab$ where a and b are numbers in the range from 2 to k inclusive. Since $P(0)$ through $P(k)$ are known to be true, a and b can each be written as a product of prime numbers. Since $k+1 = ab$, $k+1$ can also be written as a product of prime numbers. We have shown that $P(k+1)$ follows from $P(0) \wedge P(1) \wedge \cdots \wedge P(k)$, and this completes the induction. □

Exercises

1. Use induction to prove that $n^3 + 3n^2 + 2n$ is divisible by 3 for all natural numbers n.

2. Use induction to prove that

$$\sum_{i=0}^{n} r^i = \frac{1 - r^{n+1}}{1 - r}$$

for any natural number n and for any real number r such that $r \neq 1$.

3. Use induction to prove that for any natural number n,

$$\sum_{i=0}^{n} \frac{1}{2^i} = 2 - \frac{1}{2^n}$$

In addition to proving this by induction, show that it follows as a corollary of Exercise 2.

4. Use induction to prove that for any natural number n,

$$\sum_{i=0}^{n} 2^i = 2^{n+1} - 1$$

In addition to proving this by induction, show that it follows as a corollary of Exercise 2.

5. Use induction to prove that for any positive integer n,

$$\sum_{i=1}^{n} i^2 = \frac{n(n+1)(2n+1)}{6}$$

6. Use induction to prove that for any positive integer n,

$$\sum_{i=1}^{n} (2i-1) = n^2$$

7. Evaluate the following sums, using results proved in this section and in the previous exercises:

a) $1+3+5+7+9+11+13+15+17+19$

b) $1+\frac{1}{3}+\frac{1}{3^2}+\frac{1}{3^3}+\frac{1}{3^4}+\frac{1}{3^5}+\frac{1}{3^6}$

c) $50+51+52+53+\cdots+99+100$

d) $1+4+9+16+25+36+49+81+100$

e) $\frac{1}{2^2}+\frac{1}{2^3}+\cdots+\frac{1}{2^{99}}$

8. Write each of the sums in the preceding problem using summation notation.

9. Rewrite the proofs of Theorem 1.10 and Theorem 1.11 without using summation notation.

10. Use induction to prove the following generalized distributive laws for propositional logic: For any natural number $n > 1$ and any propositions q, p_1, p_2, …, p_n,

a) $q \wedge (p_1 \vee p_2 \vee \cdots \vee p_n) = (q \wedge p_1) \vee (q \wedge p_2) \vee \cdots \vee (q \wedge p_n)$

b) $q \vee (p_1 \wedge p_2 \wedge \cdots \wedge p_n) = (q \vee p_1) \wedge (q \vee p_2) \wedge \cdots \wedge (q \vee p_n)$

1.9 Application: Recursion and Induction

In computer programming, there is a technique called ***recursion*** that is closely related to induction. In a computer program, a ***subroutine*** is a named sequence of instructions for performing a certain task. When that task needs to be performed in a program, the subroutine can be ***called*** by name. A typical way to organize a program is to break down a large task into smaller, simpler subtasks by calling subroutines to perform each of the subtasks. A subroutine can perform its task by calling other subroutines to perform subtasks of the overall task. A subroutine can also call itself. That is, in the process of performing some large task, a subroutine can call itself to perform a subtask. This is known as recursion, and a subroutine that does this is said to be a ***recursive subroutine***. Recursion is appropriate when a large task can be broken into subtasks where some or all of the subtasks are smaller, simpler versions of the main task.

Like induction, recursion is often considered to be a "hard" topic by students. Professors, on the other hand, often say that they can't see what all the fuss is about, since induction and recursion are elegant methods which "obviously" work. In fairness, students have a point, since induction and recursion both manage to pull infinite rabbits out of very finite hats. But the magic is indeed elegant, and learning the trick is very worthwhile.

A simple example of a recursive subroutine is a function that computes $n!$ for a non-negative integer n. $n!$, which is read "n factorial," is defined as follows:

$$0! = 1$$

$$n! = \prod_{i=1}^{n} i \qquad \text{for } n > 0$$

For example, $5! = 1 \cdot 2 \cdot 3 \cdot 4 \cdot 5 = 120$. Note that for $n > 1$,

$$n! = \prod_{i=1}^{n} i = \left(\prod_{i=1}^{n-1} i\right) \cdot n = \big((n-1)!\big) \cdot n$$

It is also true that $n! = \big((n-1)!\big) \cdot n$ when $n = 1$. This observation makes it possible to write a recursive function to compute $n!$. (All the programming examples in this section are written in the Java programming language.)

```
int factorial( int n ) {
      // Compute n!.  Assume that n >= 0.
   int answer;
   if ( n == 0 ) {
      answer = 1;
   }
   else {
      answer = factorial( n-1 ) * n;
```

```
        }
        return answer;
    }
```

In order to compute *factorial*(n) for $n > 0$, this function first computes *factorial*$(n-1)$ by calling itself recursively. The answer from that computation is then multiplied by n to give the value of $n!$. The recursion has a base case, namely the case when $n = 0$. For the base case, the answer is computed directly rather than by using recursion. The base case prevents the recursion from continuing forever, in an infinite chain of recursive calls.

Now, as it happens, recursion is not the best way to compute $n!$. It can be computed more efficiently using a loop. Furthermore, except for small values of n, the value of $n!$ is outside the range of numbers that can be represented as 32-bit *ints*. However, ignoring these problems, the *factorial* function provides a nice first example of the interplay between recursion and induction. We can use induction to prove that *factorial*(n) does indeed compute $n!$ for $n \geq 0$. (In the proof, we pretend that the data type *int* is not limited to 32 bits. In reality, the function only gives the correct answer when the answer can be represented as a 32-bit binary number.)

Theorem 1.14. *Assume that the data type int can represent arbitrarily large integers. Under this assumption, the factorial function defined above correctly computes $n!$ for any natural number n.*

Proof. Let $P(n)$ be the statement "*factorial*(n) correctly computes $n!$." We use induction to prove that $P(n)$ is true for all natural numbers n.

Base case: In the case $n = 0$, the *if* statement in the function assigns the value 1 to the answer. Since 1 is the correct value of $0!$, *factorial*(0) correctly computes $0!$.

Inductive case: Let k be an arbitrary natural number, and assume that $P(k)$ is true. From this assumption, we must show that $P(k+1)$ is true. The assumption is that *factorial*(k) correctly computes $k!$, and we want to show that *factorial*$(k+1)$ correctly computes $(k+1)!$.

When the function computes *factorial*$(k+1)$, the value of the parameter n is $k+1$. Since $k+1 > 0$, the *if* statement in the function computes the value of *factorial*$(k+1)$ by applying the computation *factorial*$(k) * (k+1)$. We know, by the induction hypothesis, that the value computed by *factorial*(k) is $k!$. It follows that the value computed by *factorial*$(k+1)$ is $(k!)\cdot(k+1)$. As we observed above, for any $k+1 > 0$, $(k!)\cdot(k+1) = (k+1)!$. We see that *factorial*$(k+1)$ correctly computes $(k+1)!$. This completes the induction. □

In this proof, we see that the base case of the induction corresponds to the base case of the recursion, while the inductive case corresponds to a recursive subroutine call. A recursive subroutine call, like the inductive case of an induction, reduces a problem to a "simpler" or "smaller" problem, which is closer to the base case.

Another standard example of recursion is the Towers of Hanoi problem. Let n be a positive integer. Imagine a set of n disks of decreasing size, piled up in order of size, with the largest disk on the bottom and the smallest disk on top. The problem is to move this tower of disks to a second pile, following certain rules: Only one disk can be moved at a time, and a disk can only be placed on top of another disk if the disk on top is smaller. While the disks are being moved from the first pile to the second pile, disks can be kept in a third, spare pile. All the disks must at all times be in one of the three piles. For example, if there are two disks, the problem can be solved by the following sequence of moves:

```
Move disk 1 from pile 1 to pile 3
Move disk 2 from pile 1 to pile 2
Move disk 1 from pile 3 to pile 2
```

A simple recursive subroutine can be used to write out the list of moves to solve the problem for any value of n. The recursion is based on the observation that for $n > 1$, the problem can be solved as follows: Move $n - 1$ disks from pile number 1 to pile number 3 (using pile number 2 as a spare). Then move the largest disk, disk number n, from pile number 1 to pile number 2. Finally, move the $n - 1$ disks from pile number 3 to pile number 2, putting them on top of the nth disk (using pile number 1 as a spare). In both cases, the problem of moving $n - 1$ disks is a smaller version of the original problem and so can be done by recursion. Here is the subroutine, written in Java:

```
void Hanoi(int n, int A, int B, int C) {
      // List the moves for moving n disks from
      // pile number A to pile number B, using
      // pile number C as a spare.  Assume n > 0.
   if ( n == 1 ) {
      System.out.println("Move disk 1 from pile "
                        + A + " to pile " + B);
   }
   else {
      Hanoi( n-1, A, C, B );
      System.out.println("Move disk " + n
```

```
                + " from pile " + A + " to pile " + B);
            Hanoi( n-1, C, B, A );
        }
    }
```

We can use induction to prove that this subroutine does in fact solve the Towers of Hanoi problem.

Theorem 1.15. *The sequence of moves printed by the Hanoi subroutine as given above correctly solves the Towers of Hanoi problem for any integer* $n \geq 1$.

Proof. We prove by induction that whenever n is a positive integer and A, B, and C are the numbers 1, 2, and 3 in some order, the subroutine call $Hanoi(n, A, B, C)$ prints a sequence of moves that will move n disks from pile A to pile B, following all the rules of the Towers of Hanoi problem.

In the base case, $n = 1$, the subroutine call $Hanoi(1, A, B, C)$ prints out the single step "Move disk 1 from pile A to pile B," and this move does solve the problem for 1 disk.

Let k be an arbitrary positive integer, and suppose that $Hanoi(k, A, B, C)$ correctly solves the problem of moving the k disks from pile A to pile B using pile C as the spare, whenever A, B, and C are the numbers 1, 2, and 3 in some order. We need to show that $Hanoi(k+1, A, B, C)$ correctly solves the problem for $k + 1$ disks. Since $k + 1 > 1$, $Hanoi(k + 1, A, B, C)$ begins by calling $Hanoi(k, A, C, B)$. By the induction hypothesis, this correctly moves k disks from pile A to pile C. Disk number $k+1$ is not moved during this process. At that point, pile C contains the k smallest disks and pile A still contains the $(k + 1)^{st}$ disk, which has not yet been moved. So the next move printed by the subroutine, "Move disk $(k + 1)$ from pile A to pile B," is legal because pile B is empty. Finally, the subroutine calls $Hanoi(k, C, B, A)$, which, by the induction hypothesis, correctly moves the k smallest disks from pile C to pile B, putting them on top of the $(k+1)^{\text{st}}$ disk, which does not move during this process. At that point, all $(k + 1)$ disks are on pile B, so the problem for $k + 1$ disks has been correctly solved. □

Recursion is often used with linked data structures, which are data structures that are constructed by linking several objects of the same type together with pointers. (If you don't already know about objects and pointers, you will not be able to follow the rest of this section.) For an example, we'll look at the data structure known as a ***binary tree***. A binary tree consists of nodes linked together in a tree-like structure. The nodes can contain any type of data, but we will consider binary trees in which each node contains an integer. A binary tree can be empty, or it can consist of a node (called the ***root*** of the tree) and two smaller binary trees (called the ***left subtree*** and the ***right subtree*** of the tree). You can already see the recursive structure: A tree can contain smaller trees. In Java, the nodes of a tree can be represented by objects belonging to the class

```
class BinaryTreeNode {
   int item;    // An integer value stored in the node.
   BinaryTreeNode left;    // Pointer to left subtree.
   BinaryTreeNode right;   // Pointer to right subtree.
}
```

An empty tree is represented by a pointer that has the special value ***null***. If *root* is a pointer to the root node of a tree, then *root.left* is a pointer to the left subtree and *root.right* is a pointer to the right subtree. Of course, both *root.left* and *root.right* can be *null* if the corresponding subtree is empty. Similarly, *root.item* is a name for the integer in the root node.

Let's say that we want a function that will find the sum of all the integers in all the nodes of a binary tree. We can do this with a simple recursive function. The base case of the recursion is an empty tree. Since there are no integers in an empty tree, the sum of the integers in an empty tree is zero. For a non-empty tree, we can use recursion to find the sums of the integers in the left and right subtrees, and then add those sums to the integer in the root node of the tree. In Java, this can be expressed as follows:

```
int TreeSum( BinaryTreeNode root ) {
      // Find the sum of all the integers in the
      // tree that has the given root.
   int answer;
   if ( root == null ) {  // The tree is empty.
      answer = 0;
   }
   else {
      answer = TreeSum( root.left );
      answer = answer + TreeSum( root.right );
      answer = answer + root.item;
   }
   return answer;
}
```

We can use the second form of the principle of mathematical induction to prove that this function is correct.

Theorem 1.16. *The function TreeSum, defined above, correctly computes the sum of all the integers in a binary tree.*

Proof. We use induction on the number of nodes in the tree. Let $P(n)$ be the statement "*TreeSum* correctly computes the sum of the nodes in any binary tree that contains exactly n nodes." We show that $P(n)$ is true for every natural number n.

Consider the case $n = 0$. A tree with zero nodes is empty, and an empty tree is represented by a *null* pointer. In this case, the *if* statement in the definition of *TreeSum* assigns the value 0 to the answer, and this is the correct sum for an empty tree. So, $P(0)$ is true.

Let k be an arbitrary natural number, with $k > 0$. Suppose we already know $P(x)$ for each natural number x with $0 \leq x < k$. That is, *TreeSum* correctly computes the sum of all the integers in any tree that has fewer than k nodes. We must show that it follows that $P(k)$ is true, that is, that *TreeSum* works for a tree with k nodes. Suppose that *root* is a pointer to the root node of a tree that has a total of k nodes. Since the root node counts as a node, that leaves a total of $k - 1$ nodes for the left and right subtrees, so each subtree must contain fewer than k nodes. By the induction hypothesis, we know that *TreeSum*(*root.left*) correctly computes the sum of all the integers in the left subtree, and *TreeSum*(*root.right*) correctly computes the sum of all the integers in the right subtree. The sum of all the integers in the tree is *root.item* plus the sums of the integers in the subtrees, and this is the value computed by *TreeSum*. So, *TreeSum* does work for a tree with k nodes. This completes the induction. □

Note how closely the structure of the inductive proof follows the structure of the recursive function. In particular, the second principle of mathematical induction is very natural here, since the size of subtree could be anything up to one less than the size of the complete tree. It would be very difficult to use the first principle of induction in a proof about binary trees.

Exercises

1. The *Hanoi* subroutine given in this section does not just solve the Towers of Hanoi problem. It solves the problem using the minimum possible number of moves. Use induction to prove this fact.

2. Use induction to prove that the *Hanoi* subroutine uses $2^n - 1$ moves to solve the Towers of Hanoi problem for n disks. (There is a story that goes along with the Towers of Hanoi problem. It is said that on the day the world was created, a group of monks in Hanoi were set the task of solving the problem for 64 disks. They can move just one disk each day. On the day the problem

is solved, the world will end. However, we shouldn't worry too much, since $2^{64} - 1$ days is a very long time—about 50 million billion years.)

3. Consider the following recursive function:

```
int power( int x, int n ) {
      // Compute x raised to the power n.
      // Assume that n >= 0.
   int answer;
   if ( n == 0 ) {
      answer = 1;
   }
   else if (n % 2 == 0) {
      answer = power( x * x, n / 2);
   }
   else {
      answer = x * power( x * x, (n-1) / 2);
   }
   return answer;
}
```

Show that for any integer x and any non-negative integer n, the function *power*(x,n) correctly computes the value of x^n. (Assume that the *int* data type can represent arbitrarily large integers.) Note that the test "`if (n % 2 == 0)`" tests whether n is evenly divisible by 2. That is, the test is true if n is an even number. (This function is actually a very efficient way to compute x^n.)

4. A ***leaf node*** in a binary tree is a node in which both the left and the right subtrees are empty. Prove that the following recursive function correctly counts the number of leaves in a binary tree:

```
int LeafCount( BinaryTreeNode root ) {
      // Counts the number of leaf nodes in
      // the tree with the specified root.
   int count;
   if ( root == null ) {
      count = 0;
   }
   else if ( root.left == null && root.right == null ) {
      count = 1;
   }
   else {
      count = LeafCount( root.left );
      count = count + LeafCount( root.right );
   }
   return count;
}
```

5. A ***binary sort tree*** satisfies the following property: If *node* is a pointer to any node in the tree, then all the integers in the left subtree of *node* are less than *node.item* and all the integers in the right subtree of *node* are greater than or equal to *node.item*. Prove that the following recursive subroutine prints all the integers in a binary sort tree in non-decreasing order:

```
void SortPrint( BinaryTreeNode root ) {
        // Assume that root is a pointer to the
        // root node of a binary sort tree.  This
        // subroutine prints the integers in the
        // tree in non-decreasing order.
    if ( root == null ) {
           // There is nothing to print.
    }
    else {
        SortPrint( root.left );
        System.out.println( root.item );
        SortPrint( root.right );
    }
}
```

1.10 Recursive Definitions

Recursion occurs in programming when a subroutine is defined—partially, at least—in terms of itself. But recursion also occurs outside of programming. A ***recursive definition*** is a definition that includes a reference to the term that is being defined. A recursive definition defines something at least partially in terms of itself. As in the case of recursive subroutines, mathematical induction can often be used to prove facts about things that are defined recursively.

As already noted, there is a recursive definition for $n!$, for n in $\mathbb{N}$. We can define $0! = 1$ and $n! = n\cdot(n-1)!$ for $n > 0$. Other sequences of numbers can also be defined recursively. For example, the famous ***Fibonacci sequence*** is the sequence of numbers f_0, f_1, f_2, ..., defined recursively by

$$\begin{aligned} f_0 &= 0 \\ f_1 &= 1 \\ f_n &= f_{n-1} + f_{n-2} \qquad \text{for } n > 1 \end{aligned}$$

Using this definition, we compute that

$$\begin{aligned} f_2 &= f_1 + f_0 = 0 + 1 = 1 \\ f_3 &= f_2 + f_1 = 1 + 1 = 2 \\ f_4 &= f_3 + f_2 = 2 + 1 = 3 \\ f_5 &= f_4 + f_3 = 3 + 2 = 5 \\ f_6 &= f_5 + f_4 = 5 + 3 = 8 \\ f_7 &= f_6 + f_5 = 8 + 5 = 13 \end{aligned}$$

and so on. Based on this definition, we can use induction to prove facts about the Fibonacci sequence. We can prove, for example, that f_n grows exponentially with n, even without finding an exact formula for f_n:

Theorem 1.17. *The Fibonacci sequence, f_0, f_1, f_2, ..., satisfies $f_n > \left(\frac{3}{2}\right)^{n-1}$, for $n \geq 6$.*

Proof. We prove this by induction on n. For $n = 6$, we have that $f_n = 8$ while $1.5^{n-1} = 1.5^5$, which is about 7.6. So $f_n > 1.5^{n-1}$ for $n = 6$. Similarly, for $n = 7$, we have $f_n = 13$ and $1.5^{n-1} = 1.5^6$, which is about 11.4. So $f_n > 1.5^{n-1}$ for $n = 7$.

Now suppose that k is an arbitrary integer with $k > 7$. Suppose that we already know that $f_n > 1.5^{n-1}$ for $n = k - 1$ and for $n = k - 2$. We want to show that the inequality then holds for $n = k$ as well. But

$$\begin{aligned} f_k &= f_{k-1} + f_{k-2} \\ &> 1.5^{(k-1)-1} + 1.5^{(k-2)-1} && \text{(by the induction hypothesis)} \\ &= 1.5^{k-2} + 1.5^{k-3} \\ &= (1.5) \cdot (1.5^{k-3}) + (1.5^{k-3}) \\ &= (2.5) \cdot (1.5^{k-3}) \\ &> (1.5^2) \cdot (1.5^{k-3}) && \text{(since } 1.5^2 = 2.25) \\ &= 1.5^{k-1} \end{aligned}$$

This string of equalities and inequalities shows that $f_k > 1.5^{k-1}$. This completes the induction and proves the theorem. □

Exercises

1. Prove that the Fibonacci sequence, f_0, f_1, f_2, ..., satisfies $f_n < 2^n$ for all natural numbers n.

2. Suppose that a_1, a_2, a_3, ..., is a sequence of numbers which is defined recursively by $a_1 = 1$ and $a_n = 2a_{n-1} + 2^{n-1}$ for $n > 1$. Prove that $a_n = n2^{n-1}$ for every positive integer n.

Chapter 2

Sets, Functions, and Relations

WE DEAL WITH THE COMPLEXITY of the world by putting things into categories. There are not just hordes of individual creatures. There are dogs, cats, elephants, and mice. There are mammals, insects, and fish. Animals, vegetables, and minerals. Solids, liquids, and gases. Things that are red. Big cities. Pleasant memories.... Categories build on categories. They are the subject and the substance of thought.

In mathematics, which operates in its own abstract and rigorous world, categories are modeled by ***sets***. A set is just a collection of elements. Along with logic, sets form the "foundation" of mathematics, just as categories are part of the foundation of day-to-day thought. In this chapter, we study sets and relationships among sets.

2.1 Basic Concepts

A ***set*** is a collection of ***elements***. A set is defined entirely by the elements that it contains. An element can be anything, including another set. You will notice that this is not a precise mathematical definition. Instead, it is an intuitive description of what the word "set" is supposed to mean: Any time you have a bunch of entities and you consider them as a unit, you have a set. Mathematically, sets are really defined by the operations that can be performed on them. These operations model things that can be done with collections of objects in the real world. These operations are the subject of the branch of mathematics known as ***set theory***.

The most basic operation in set theory is forming a set from a given list of specific entities. The set that is formed in this way is denoted by enclosing the list of entities between a left brace, "{", and a right brace, "}". The entities in the list are separated by commas. For example, the set denoted by

$$\{\ 17,\ \pi,\ \texttt{New York City},\ \texttt{Barack Obama},\ \texttt{Big Ben}\ \}$$

is the set that contains the entities 17, π, New York City, Barack Obama, and Big Ben. These entities are the elements of the set. Since we assume that a set is completely defined by the elements that it contains, the set is well-defined. Of course, we still haven't said what it means to be an "entity." Something as definite as "New York City" should qualify, except that it doesn't seem like New York City really belongs in the world of Mathematics. The problem is that mathematics is supposed to be its own self-contained world, but it is supposed to model the real world. When we use mathematics to model the real world, we admit entities such as New York City and even Big Ben. But when we are doing mathematics *per se*, we'll generally stick to obviously mathematical entities such as the integer 17 or the real number π. We will also use letters such as a and b to refer to entities. For example, when I say something like "Let A be the set $\{a, b, c\}$," I mean a, b, and c to be particular, but unspecified, entities.

It's important to understand that a set is defined by the elements that it contains, and not by the order in which those elements might be listed. For example, the notations $\{a, b, c, d\}$ and $\{b, c, a, d\}$ define the same set. Furthermore, a set can only contain one copy of a given element, even if the notation that specifies the set lists the element twice. This means that $\{a, b, a, a, b, c, a\}$ and $\{a, b, c\}$ specify exactly the same set. Note in particular that it's incorrect to say that the set $\{a, b, a, a, b, c, a\}$ contains seven elements, since some of the elements in the list are identical. The notation $\{a, b, c\}$ can lead to some confusion, since it might not be clear whether the letters a, b, and c are assumed to refer to three *different* entities. A mathematician would generally *not* make this assumption without stating it explicitly, so that the set denoted by $\{a, b, c\}$ could actually contain either one, two, or three elements. When it is important that different letters refer to different entities, I will say so explicitly, as in "Consider the set $\{a, b, c\}$, where a, b, and c are distinct."

The symbol $\in$ is used to express the relation "is an element of." That is, if a is an entity and A is a set, then $a \in A$ is a statement that is true if and only if a is one of the elements of A. In that case, we also say that a is a ***member*** of the set A. The assertion that a is not an element of A is expressed by the notation $a \not\in A$. Note that both $a \in A$ and $a \not\in A$ are statements in the sense of propositional logic. That is, they are assertions which can be either true or false. The statement $a \not\in A$ is equivalent to

$\neg(a \in A)$.

It is possible for a set to be empty, that is, to contain no elements whatsoever. Since a set is completely determined by the elements that it contains, there is only one set that contains no elements. This set is called the ***empty set***, and it is denoted by the symbol $\emptyset$. Note that for any element a, the statement $a \in \emptyset$ is false. The empty set, $\emptyset$, can also be denoted by an empty pair of braces, $\{\ \}$.

If A and B are sets, then, by definition, A is equal to B if and only if they contain exactly the same elements. In this case, we write $A = B$. Using the notation of predicate logic, we can say that $A = B$ if and only if $\forall x(x \in A \leftrightarrow x \in B)$.

Suppose now that A and B are sets such that every element of A is an element of B. In that case, we say that A is a ***subset*** of B, i.e. A is a subset of B if and only if $\forall x(x \in A \to x \in B)$. The fact that A is a subset of B is denoted by $A \subseteq B$. Note that $\emptyset$ is a subset of every set B: $x \in \emptyset$ is false for any x, and so given any B, $(x \in \emptyset \to x \in B)$ is true for all x.

If $A = B$, then it is automatically true that $A \subseteq B$ and that $B \subseteq A$. The converse is also true: If $A \subseteq B$ and $B \subseteq A$, then $A = B$. This follows from the fact that for any x, the statement $(x \in A \leftrightarrow x \in B)$ is logically equivalent to the statement $(x \in A \to x \in B) \wedge (x \in B \to x \in A)$. This fact is important enough to state as a theorem.

Theorem 2.1. *Let A and B be sets. Then $A = B$ if and only if both $A \subseteq B$ and $B \subseteq A$.*

This theorem expresses the following advice: If you want to check that two sets, A and B, are equal, you can do so in two steps. First check that every element of A is also an element of B, and then check that every element of B is also an element of A.

If $A \subseteq B$ but $A \neq B$, we say that A is a ***proper subset*** of B. We use the notation $A \subsetneq B$ to mean that A is a proper subset of B. That is, $A \subsetneq B$ if and only if $A \subseteq B \wedge A \neq B$. We will sometimes use $A \supseteq B$ as an equivalent notation for $B \subseteq A$, and $A \supsetneq B$ as an equivalent for $B \subsetneq A$.

A set can contain an infinite number of elements. In such a case, it is not possible to list all the elements in the set. Sometimes the ellipsis "..." is used to indicate a list that continues on infinitely. For example, $\mathbb{N}$, the set of natural numbers, can be specified as

$$\mathbb{N} = \{0, 1, 2, 3, \dots\}$$

However, this is an informal notation, which is not really well-defined, and it should only be used in cases where it is clear what it means. It's not very useful to say that "the set of prime numbers is $\{2, 3, 5, 7, 11, 13, \dots\}$," and it is completely meaningless to talk about "the set $\{17, 42, 105, \dots\}$." Clearly, we need another way to specify sets besides listing their elements. The need is fulfilled by predicates.

If $P(x)$ is a predicate, then we can form the set that contains all entities a such that a is in the domain of discourse for P and $P(a)$ is true. The notation $\{x \mid P(x)\}$ is used to denote this set. The name of the variable, x, is arbitrary, so the same set could equally well be denoted as $\{z \mid P(z)\}$ or $\{r \mid P(r)\}$. The notation $\{x \mid P(x)\}$ can be read "the set of x such that $P(x)$." For example, if $E(x)$ is the predicate "x is an even number," and if the domain of discourse for E is the set $\mathbb{N}$ of natural numbers, then the notation $\{x \mid E(x)\}$ specifies the set of even natural numbers. That is,

$$\{x \mid E(x)\} = \{0, 2, 4, 6, 8, \dots\}$$

It turns out, for deep and surprising reasons that we will discuss later in this section, that we have to be a little careful about what counts as a predicate. In order for the notation $\{x \mid P(x)\}$ to be valid, we have to assume that the domain of discourse of P is in fact a set. (You might wonder how it could be anything else. That's the surprise!) Often, it is useful to specify the domain of discourse explicitly in the notation that defines a set. In the above example, to make it clear that x must be a natural number, we could write the set as $\{x \in \mathbb{N} \mid E(x)\}$. This notation can be read as "the set of all x in $\mathbb{N}$ such that $E(x)$." More generally, if X is a set and P is a predicate whose domain of discourse includes all the elements of X, then the notation

$$\{x \in X \mid P(x)\}$$

is the set that consists of all entities a that are members of the set X and for which $P(a)$ is true. In this notation, we don't have to assume that the domain of discourse for P is a set, since we are effectively limiting the domain of discourse to the set X. The set denoted by $\{x \in X \mid P(x)\}$ could also be written as $\{x \mid x \in X \wedge P(x)\}$.

We can use this notation to define the set of prime numbers in a rigorous way. A prime number is a natural number n which is greater than 1 and which satisfies the property that for any factorization $n = xy$, where x and y are natural numbers, either x or y must be n. We can express this definition as a predicate and define the set of prime numbers as

$$\{n \in \mathbb{N} \mid (n > 1) \wedge \forall x \forall y \big((x \in \mathbb{N} \wedge y \in \mathbb{N} \wedge n = xy) \rightarrow (x = n \vee y = n)\big)\}.$$

Admittedly, this definition is hard to take in in one gulp. But this example shows that it is possible to define complex sets using predicates.

Now that we have a way to express a wide variety of sets, we turn to operations that can be performed on sets. The most basic operations on sets are ***union*** and ***intersection***. If A and B are sets, then we define the union of A and B to be the set that contains all the elements of A together with all the elements of B. The union of A and B is denoted by $A \cup B$. The union can be defined formally as

$$A \cup B = \{x \,|\, x \in A \vee x \in B\}.$$

The intersection of A and B is defined to be the set that contains every entity that is both a member of A and a member of B. The intersection of A and B is denoted by $A \cap B$. Formally,

$$A \cap B = \{x \,|\, x \in A \wedge x \in B\}.$$

An entity gets into $A \cup B$ if it is in *either* A or B. It gets into $A \cap B$ if it is in *both* A and B. Note that the symbol for the logical "or" operator, $\vee$, is similar to the symbol for the union operator, $\cup$, while the logical "and" operator, $\wedge$, is similar to the intersection operator, $\cap$.

The ***set difference*** of two sets, A and B, is defined to be the set of all entities that are members of A but are not members of B. The set difference of A and B is denoted $A \smallsetminus B$. The idea is that $A \smallsetminus B$ is formed by starting with A and then removing any element that is also found in B. Formally,

$$A \smallsetminus B = \{x \,|\, x \in A \wedge x \notin B\}.$$

Union and intersection are clearly commutative operations. That is, $A \cup B = B \cup A$ and $A \cap B = B \cap A$ for any sets A and B. However, set difference is not commutative. In general, $A \smallsetminus B \neq B \smallsetminus A$.

Suppose that $A = \{a, b, c\}$, that $B = \{b, d\}$, and that $C = \{d, e, f\}$. Then we can apply the definitions of union, intersection, and set difference to compute, for example, that:

$$\begin{array}{lll} A \cup B = \{a, b, c, d\} & A \cap B = \{b\} & A \smallsetminus B = \{a, c\} \\ A \cup C = \{a, b, c, d, e, f\} & A \cap C = \emptyset & A \smallsetminus C = \{a, b, c\} \end{array}$$

In this example, the sets A and C have no elements in common, so that $A \cap C = \emptyset$. There is a term for this: Two sets are said to be ***disjoint*** if they have no elements in common. That is, for any sets A and B, A and B are said to be disjoint if and only if $A \cap B = \emptyset$.

Of course, the set operations can also be applied to sets that are defined by predicates. For example, let $L(x)$ be the predicate "x is lucky," and let $W(x)$ be the predicate "x is wise," where the domain of discourse for each

Notation	Definition
$a \in A$	a is a member (or element) of A
$a \notin A$	$\neg(a \in A)$, a is not a member of A
$\emptyset$	the empty set, which contains no elements
$A \subseteq B$	A is a subset of B, $\forall x(x \in A \rightarrow x \in B)$
$A \subsetneq B$	A is a proper subset of B, $A \subseteq B \wedge A \neq B$
$A \supseteq B$	A is a superset of B, same as $B \subseteq A$
$A \supsetneq B$	A is a proper superset of B, same as $B \supsetneq A$
$A = B$	A and B have the same members, $A \subseteq B \wedge B \subseteq A$
$A \cup B$	union of A and B, $\{x \mid x \in A \vee x \in B\}$
$A \cap B$	intersection of A and B, $\{x \mid x \in A \wedge x \in B\}$
$A \smallsetminus B$	set difference of A and B, $\{x \mid x \in A \wedge x \notin B\}$
$\mathcal{P}(A)$	power set of A, $\{X \mid X \subseteq A\}$

Figure 2.1: Some of the notations that are defined in this section. A and B are sets, and a is an entity.

predicate is the set of people. Let $X = \{x \mid L(x)\}$, and let $Y = \{x \mid W(x)\}$. Then

$$\begin{aligned}
X \cup Y &= \{x \mid L(x) \vee W(x)\} = \{\text{people who are lucky or wise}\} \\
X \cap Y &= \{x \mid L(x) \wedge W(x)\} = \{\text{people who are lucky and wise}\} \\
X \smallsetminus Y &= \{x \mid L(x) \wedge \neg W(x)\} = \{\text{people who are lucky but not wise}\} \\
Y \smallsetminus X &= \{x \mid W(x) \wedge \neg L(x)\} = \{\text{people who are wise but not lucky}\}
\end{aligned}$$

You have to be a little careful with the English word "and." We might say that the set $X \cup Y$ contains people who are lucky *and* people who are wise. But what this means is that a person gets into the set $X \cup Y$ either by being lucky *or* by being wise, so $X \cup Y$ is defined using the logical "or" operator, $\vee$.

Sets can contain other sets as elements. For example, the notation $\{a, \{b\}\}$ defines a set that contains two elements, the entity a and the set $\{b\}$. Since the set $\{b\}$ is a member of the set $\{a, \{b\}\}$, we have that $\{b\} \in \{a, \{b\}\}$. On the other hand, provided that $a \neq b$, the statement $\{b\} \subseteq \{a, \{b\}\}$ is false, since saying $\{b\} \subseteq \{a, \{b\}\}$ is equivalent to saying that $b \in \{a, \{b\}\}$, and the entity b is not one of the two members of $\{a, \{b\}\}$. For the entity a, it is true that $\{a\} \subseteq \{a, \{b\}\}$.

Given a set A, we can construct the set that contains all the subsets of A. This set is called the **power set** of A, and is denoted $\mathcal{P}(A)$. Formally, we define

$$\mathcal{P}(A) = \{X \mid X \subseteq A\}.$$

For example, if $A = \{a, b\}$, then the subsets of A are the empty set, $\{a\}$, $\{b\}$, and $\{a, b\}$, so the power set of A is set given by

$$\mathcal{P}(A) = \{\, \emptyset,\ \{a\},\ \{b\},\ \{a, b\}\, \}.$$

Note that since the empty set is a *subset* of any set, the empty set is an *element* of the power set of any set. That is, for any set A, $\emptyset \subseteq A$ and $\emptyset \in \mathcal{P}(A)$. Since the empty set is a subset of itself, and is its only subset, we have that $\mathcal{P}(\emptyset) = \{\emptyset\}$. The set $\{\emptyset\}$ is not empty. It contains one element, namely $\emptyset$.

We remarked earlier in this section that the notation $\{x \mid P(x)\}$ is only valid if the domain of discourse of P is a set. This might seem a rather puzzling thing to say—after all, why and how would the domain of discourse be anything else? The answer is related to Russell's Paradox, which we mentioned briefly in Chapter 1 and which shows that it is logically impossible for the set of all sets to exist. This impossibility can be demonstrated using a proof by contradiction. In the proof, we use the existence of the set of all sets to define another set which cannot exist because its existence would lead to a logical contradiction.

Theorem 2.2. *There is no set of all sets.*

Proof. Suppose that the set of all sets exists. We will show that this assumption leads to a contradiction. Let V be the set of all sets. We can then define the set R to be the set which contains every set that does not contain itself. That is,

$$R = \{X \in V \mid X \notin X\}$$

Now, we must have either $R \in R$ or $R \notin R$. We will show that either case leads to a contradiction.

Consider the case where $R \in R$. Since $R \in R$, R must satisfy the condition for membership in R. A set X is in R iff $X \notin X$. To say that R satisfies this condition means that $R \notin R$. That is, from the fact that $R \in R$, we deduce the contradiction that $R \notin R$.

Now consider the remaining case, where $R \notin R$. Since $R \notin R$, R does not satisfy the condition for membership in R. Since the condition for membership is that $R \notin R$, and this condition is false, the statement $R \notin R$ must be false. But this means that the statement $R \in R$ is true. From the fact that $R \notin R$, we deduce the contradiction that $R \in R$.

Since both possible cases, $R \in R$ and $R \notin R$, lead to contradictions, we see that it is not possible for R to exist. Since the existence of R follows from the existence of V, we see that V also cannot exist. □

To avoid Russell's paradox, we must put limitations on the construction of new sets. We can't force the set of all sets into existence simply by thinking of it. We can't form the set $\{x \mid P(x)\}$ unless the domain of discourse of P is a set. Any predicate Q can be used to form a set $\{x \in X \mid Q(x)\}$, but this notation requires a pre-existing set X. Predicates can be used to form subsets of existing sets, but they can't be used to form new sets completely from scratch.

The notation $\{x \in A \mid P(x)\}$ is a convenient way to effectively limit the domain of discourse of a predicate, P, to members of a set, A, that we are actually interested in. We will use a similar notation with the quantifiers $\forall$ and $\exists$. The proposition $(\forall x \in A)(P(x))$ is true if and only if $P(a)$ is true for every element a of the set A. And the proposition $(\exists x \in A)(P(x))$ is true if and only if there is some element a of the set A for which $P(a)$ is true. These notations are valid only when A is contained in the domain of discourse for P. As usual, we can leave out parentheses when doing so introduces no ambiguity. So, for example, we might write $\forall x \in A\ P(x)$.

We end this section with proofs of the two forms of the principle of mathematical induction. These proofs were omitted from the previous chapter, but only for the lack of a bit of set notation. In fact, the principle of mathematical induction is valid only because it follows from one of the basic axioms that define the natural numbers, namely the fact that any non-empty set of natural numbers has a smallest element. Given this axiom, we can use it to prove the following two theorems:

Theorem 2.3. *Let P be a one-place predicate whose domain of discourse includes the natural numbers. Suppose that $P(0) \wedge \big(\forall k \in \mathbb{N}\ (P(k) \to P(k+1))\big)$. Then $\forall n \in \mathbb{N}$, $P(n)$.*

Proof. Suppose that both $P(0)$ and $\big(\forall k \in \mathbb{N}\,(P(k) \to P(k+1))\big)$ are true, but that $\big(\forall n \in \mathbb{N},\ P(n)\big)$ is false. We show that this assumption leads to a contradiction.

Since the statement $\forall n \in \mathbb{N},\ P(n)$ is false, its negation, $\neg(\forall n \in \mathbb{N},\ P(n))$, is true. The negation is equivalent to $\exists n \in \mathbb{N},\ \neg P(n)$. Let $X = \{n \in \mathbb{N} \mid \neg P(n)\}$. Since $\exists n \in \mathbb{N},\ \neg P(n)$ is true, we know that X is not empty. Since X is a non-empty set of natural numbers, it has a smallest element. Let x be the smallest element of X. That is, x is the smallest natural number such that $P(x)$ is false. Since we know that $P(0)$ is true, x cannot be 0. Let $y = x - 1$. Since $x \neq 0$, y is a natural number. Since $y < x$, we know, by the definition of x, that $P(y)$ is true. We also know that $\forall k \in \mathbb{N}\,(P(k) \rightarrow P(k+1))$ is true. In particular, taking $k = y$, we know that $P(y) \rightarrow P(y+1)$. Since $P(y)$ and $P(y) \rightarrow P(y+1)$, we deduce by *modus ponens* that $P(y+1)$ is true. But $y+1 = x$, so we have deduced that $P(x)$ is true. This contradicts the fact that $P(x)$ is false. This contradiction proves the theorem. □

Theorem 2.4. *Let P be a one-place predicate whose domain of discourse includes the natural numbers. Suppose that $P(0)$ is true and that*

$$(P(0) \wedge P(1) \wedge \cdots \wedge P(k)) \rightarrow P(k+1)$$

is true for each natural number $k \geq 0$. Then it is true that $\forall n \in \mathbb{N},\ P(n)$.

Proof. Suppose that P is a predicate that satisfies the hypotheses of the theorem, and suppose that the statement $\forall n \in \mathbb{N},\ P(n)$ is false. We show that this assumption leads to a contradiction.

Let $X = \{n \in \mathbb{N} \mid \neg P(n)\}$. Because of the assumption that $\forall n \in \mathbb{N},\ P(n)$ is false, X is non-empty. It follows that X has a smallest element. Let x be the smallest element of X. The assumption that $P(0)$ is true means that $0 \notin X$, so we must have $x > 0$. Since x is the smallest natural number for which $P(x)$ is false, we know that $P(0)$, $P(1)$, …, and $P(x-1)$ are all true. From this and the fact that $(P(0) \wedge P(1) \wedge \cdots \wedge P(x-1)) \rightarrow P(x)$, we deduce that $P(x)$ is true. But this contradicts the fact that $P(x)$ is false. This contradiction proves the theorem. □

Exercises

1. If we don't make the assumption that a, b, and c are distinct, then the set denoted by $\{a, b, c\}$ might actually contain either 1, 2, or 3 elements. How many different elements might the set $\{\, a,\ b,\ \{a\},\ \{a, c\},\ \{a, b, c\}\,\}$ contain? Explain your answer.

2. Compute $A \cup B$, $A \cap B$, and $A \smallsetminus B$ for each of the following pairs of sets

a) $A = \{a, b, c\},\ B = \emptyset$

b) $A = \{1, 2, 3, 4, 5\},\ B = \{2, 4, 6, 8, 10\}$

c) $A = \{a, b\}$, $B = \{a, b, c, d\}$
d) $A = \{a, b, \{a, b\}\}$, $B = \{\{a\}, \{a, b\}\}$

3. Recall that $\mathbb{N}$ represents the set of natural numbers. That is, $\mathbb{N} = \{0, 1, 2, 3, \ldots\}$. Let $X = \{n \in \mathbb{N} \,|\, n \geq 5\}$, let $Y = \{n \in \mathbb{N} \,|\, n \leq 10\}$, and let $Z = \{n \in \mathbb{N} \,|\, n \text{ is an even number}\}$. Find each of the following sets:
a) $X \cap Y$ **b)** $X \cup Y$ **c)** $X \smallsetminus Y$ **d)** $\mathbb{N} \smallsetminus Z$
e) $X \cap Z$ **f)** $Y \cap Z$ **g)** $Y \cup Z$ **h)** $Z \smallsetminus \mathbb{N}$

4. Find $\mathcal{P}(\{1, 2, 3\})$. (It has eight elements.)

5. Assume that a and b are entities and that $a \neq b$. Let A and B be the sets defined by $A = \{\, a, \{b\}, \{a, b\} \,\}$ and $B = \{\, a, b, \{a, \{b\}\} \,\}$. Determine whether each of the following statements is true or false. Explain your answers.
a) $b \in A$ **b)** $\{a, b\} \subseteq A$ **c)** $\{a, b\} \subseteq B$
d) $\{a, b\} \in B$ **e)** $\{a, \{b\}\} \in A$ **f)** $\{a, \{b\}\} \in B$

6. Since $\mathcal{P}(A)$ is a set, it is possible to form the set $\mathcal{P}\big(\mathcal{P}(A)\big)$. What is $\mathcal{P}\big(\mathcal{P}(\emptyset)\big)$? What is $\mathcal{P}\big(\mathcal{P}(\{a, b\})\big)$? (It has sixteen elements.)

7. In the English sentence, "She likes men who are tall, dark, and handsome," does she like an intersection or a union of sets of men? How about in the sentence, "She likes men who are tall, men who are dark, and men who are handsome?"

8. If A is any set, what can you say about $A \cup A$? About $A \cap A$? About $A \smallsetminus A$? Why?

9. Suppose that A and B are sets such that $A \subseteq B$. What can you say about $A \cup B$? About $A \cap B$? About $A \smallsetminus B$? Why?

10. Suppose that A, B, and C are sets. Show that $C \subseteq A \cap B$ if and only if $(C \subseteq A) \wedge (C \subseteq B)$.

11. Suppose that A, B, and C are sets, and that $A \subseteq B$ and $B \subseteq C$. Show that $A \subseteq C$.

12. Suppose that A and B are sets such that $A \subseteq B$. Is it necessarily true that $\mathcal{P}(A) \subseteq \mathcal{P}(B)$? Why or why not?

13. Let M be any natural number, and let $P(n)$ be a predicate whose domain of discourse includes all natural numbers greater than or equal to M. Suppose that $P(M)$ is true, and suppose that $P(k) \rightarrow P(k+1)$ for all $k \geq M$. Show that $P(n)$ is true for all $n \geq M$.

2.2 The Boolean Algebra of Sets

It is clear that set theory is closely related to logic. The intersection and union of sets can be defined in terms of the logical "and" and logical "or" operators. The notation $\{x \,|\, P(x)\}$ makes it possible to use predicates to specify sets. And if A is any set, then the formula $x \in A$ defines a one place predicate that is true for an entity x if and only if x is a member of A. So it should not be a surprise that many of the rules of logic have analogs in set theory.

For example, we have already noted that $\cup$ and $\cap$ are commutative operations. This fact can be verified using the rules of logic. Let A and B be sets. According to the definition of equality of sets, we can show that $A \cup B = B \cup A$ by showing that $\forall x \big((x \in A \cup B) \leftrightarrow (x \in B \cup A)\big)$. But for any x,

$$\begin{aligned} x \in A \cup B &\leftrightarrow x \in A \vee x \in B && \text{(definition of } \cup\text{)} \\ &\leftrightarrow x \in B \vee x \in A && \text{(commutativity of } \vee\text{)} \\ &\leftrightarrow x \in B \cup A && \text{(definition of } \cup\text{)} \end{aligned}$$

The commutativity of $\cap$ follows in the same way from the definition of $\cap$ in terms of $\wedge$ and the commutativity of $\wedge$, and a similar argument shows that union and intersection are associative operations.

The distributive laws for propositional logic give rise to two similar rules in set theory. Let A, B, and C be any sets. Then

$$A \cup (B \cap C) = (A \cup B) \cap (A \cup C)$$

and

$$A \cap (B \cup C) = (A \cap B) \cup (A \cap C)$$

These rules are called the distributive laws for set theory. To verify the first of these laws, we just have to note that for any x,

$$\begin{aligned} & x \in A \cup (B \cap C) \\ &\quad \leftrightarrow (x \in A) \vee ((x \in B) \wedge (x \in C)) && \text{(definition of } \cup, \cap\text{)} \\ &\quad \leftrightarrow ((x \in A) \vee (x \in B)) \wedge ((x \in A) \vee (x \in C)) && \text{(distributivity of } \vee\text{)} \\ &\quad \leftrightarrow (x \in A \cup B) \wedge (x \in A \cup C) && \text{(definition of } \cup\text{)} \\ &\quad \leftrightarrow x \in ((A \cup B) \cap (A \cup C)) && \text{(definition of } \cap\text{)} \end{aligned}$$

The second distributive law for sets follows in exactly the same way.

While $\cup$ is analogous to $\vee$ and $\cap$ is analogous to $\wedge$, we have not yet seen any operation is set theory that is analogous to the logical "not" operator, $\neg$. Given a set A, it is tempting to try to define $\{x \mid \neg(x \in A)\}$, the set that contains everything that does not belong to A. Unfortunately, the rules of set theory do not allow us to define such a set. The notation $\{x \mid P(x)\}$ can only be used when the domain of discourse of P is a set, so there must be an underlying set from which the elements that are/are not in A are chosen, i.e. some underlying set of which A is a subset. We can get around this problem by restricting the discussion to subsets of some fixed set. This set will be known as the ***universal set***. Keep in mind that the universal set is only universal for some particular discussion. It is simply some set that is large enough to contain all the sets under discussion as subsets. Given a universal set U and any subset A of U, we can define the set $\{x \in U \mid \neg(x \in A)\}$.

Definition 2.1. Let U be a given universal set, and let A be any subset of U. We define the ***complement*** of A in U to be the set $\overline{A}$ that is defined by $\overline{A} = \{x \in U \mid x \notin A\}$.

Usually, we will refer to the complement of A in U simply as the complement of A, but you should remember that whenever complements of sets are used, there must be some universal set in the background.

Given the complement operation on sets, we can look for analogs to the rules of logic that involve negation. For example, we know that $p \wedge \neg p = \mathbb{F}$ for any proposition p. It follows that for any subset A of U,

$$\begin{aligned} A \cap \overline{A} &= \{x \in U \mid (x \in A) \wedge (x \in \overline{A})\} && \text{(definition of } \cap\text{)} \\ &= \{x \in U \mid (x \in A) \wedge (x \notin A)\} && \text{(definition of complement)} \\ &= \{x \in U \mid (x \in A) \wedge \neg(x \in A)\} && \text{(definition of } \notin\text{)} \\ &= \emptyset \end{aligned}$$

the last equality following because the proposition $(x \in A) \wedge \neg(x \in A)$ is false for any x. Similarly, we can show that $A \cup \overline{A} = U$ and that $\overline{\overline{A}} = A$ (where $\overline{\overline{A}}$ is the complement of the complement of A, that is, the set obtained by taking the complement of $\overline{A}$.)

The most important laws for working with complements of sets are DeMorgan's Laws for sets. These laws, which follow directly from DeMorgan's Laws for logic, state that for any subsets A and B of a universal set U,

$$\overline{A \cup B} = \overline{A} \cap \overline{B}$$

and

$$\overline{A \cap B} = \overline{A} \cup \overline{B}$$

Double complement	$\overline{\overline{A}} = A$
Miscellaneous laws	$A \cup \overline{A} = U$ $A \cap \overline{A} = \emptyset$ $\emptyset \cup A = A$ $\emptyset \cap A = \emptyset$
Idempotent laws	$A \cap A = A$ $A \cup A = A$
Commutative laws	$A \cap B = B \cap A$ $A \cup B = B \cup A$
Associative laws	$A \cap (B \cap C) = (A \cap B) \cap C$ $A \cup (B \cup C) = (A \cup B) \cup C$
Distributive laws	$A \cap (B \cup C) = (A \cap B) \cup (A \cap C)$ $A \cup (B \cap C) = (A \cup B) \cap (A \cup C)$
DeMorgan's laws	$\overline{A \cap B} = \overline{A} \cup \overline{B}$ $\overline{A \cup B} = \overline{A} \cap \overline{B}$

Figure 2.2: Some Laws of Boolean Algebra for sets. A, B, and C are sets. For the laws that involve the complement operator, they are assumed to be subsets of some universal set, U. For the most part, these laws correspond directly to laws of Boolean Algebra for propositional logic as given in Figure 1.2.

For example, we can verify the first of these laws with the calculation

$$\begin{aligned}
\overline{A \cup B} &= \{x \in U \mid x \notin (A \cup B)\} && \text{(definition of complement)}\\
&= \{x \in U \mid \neg(x \in A \cup B)\} && \text{(definition of } \notin)\\
&= \{x \in U \mid \neg(x \in A \vee x \in B)\} && \text{(definition of } \cup)\\
&= \{x \in U \mid (\neg(x \in A)) \wedge (\neg(x \in B))\} && \text{(DeMorgan's Law for logic)}\\
&= \{x \in U \mid (x \notin A) \wedge (x \notin B)\} && \text{(definition of } \notin)\\
&= \{x \in U \mid (x \in \overline{A}) \wedge (x \in \overline{B})\} && \text{(definition of complement)}\\
&= \overline{A} \cap \overline{B} && \text{(definition of } \cap)
\end{aligned}$$

An easy inductive proof can be used to verify generalized versions of DeMorgan's Laws for set theory. (In this context, all sets are assumed to be subsets of some unnamed universal set.) A simple calculation verifies DeMorgan's Law for three sets:

$$\begin{aligned}
\overline{A \cup B \cup C} &= \overline{(A \cup B) \cup C} \\
&= \overline{(A \cup B)} \cap \overline{C} && \text{(by DeMorgan's Law for two sets)} \\
&= (\overline{A} \cap \overline{B}) \cap \overline{C} && \text{(by DeMorgan's Law for two sets)} \\
&= \overline{A} \cap \overline{B} \cap \overline{C}
\end{aligned}$$

From there, we can derive similar laws for four sets, five sets, and so on. However, just saying "and so on" is not a rigorous proof of this fact. Here is a rigorous inductive proof of a generalized DeMorgan's Law:

Theorem 2.5. *For any natural number $n \geq 2$ and for any sets X_1, X_2, ..., X_n,*

$$\overline{X_1 \cup X_2 \cup \cdots \cup X_n} = \overline{X_1} \cap \overline{X_2} \cap \cdots \cap \overline{X_2}$$

Proof. We give a proof by induction. In the base case, $n = 2$, the statement is that $\overline{X_1 \cup X_2} = \overline{X_1} \cap \overline{X_n}$. This is true since it is just an application of DeMorgan's law for two sets.

For the inductive case, suppose that the statement is true for $n = k$. We want to show that it is true for $n = k + 1$. Let X_1, X_2, ..., X_{k+1} be any k sets. Then we have:

$$\begin{aligned}
\overline{X_1 \cup X_2 \cup \cdots \cup X_{k+1}} &= \overline{(X_1 \cup X_2 \cup \cdots \cup X_k) \cup X_{k+1}} \\
&= \overline{(X_1 \cup X_2 \cup \cdots \cup X_k)} \cap \overline{X_{k+1}} \\
&= (\overline{X_1} \cap \overline{X_2} \cap \cdots \cap \overline{X_k}) \cap \overline{X_{k+1}} \\
&= \overline{X_1} \cap \overline{X_2} \cap \cdots \cap \overline{X_{k+1}}
\end{aligned}$$

In this computation, the second step follows by DeMorgan's Law for two sets, while the third step follows from the induction hypothesis. □

Just as the laws of logic allow us to do algebra with logical formulas, the laws of set theory allow us to do algebra with sets. Because of the close relationship between logic and set theory, their algebras are very similar. The algebra of sets, like the algebra of logic, is Boolean algebra. When George Boole wrote his 1854 book about logic, it was really as much about set theory as logic. In fact, Boole did not make a clear distinction between a predicate and the set of objects for which that predicate is true. His algebraic laws and formulas apply equally to both cases. More exactly, if we consider only subsets of some given universal set U, then there is a direct correspondence between the basic symbols and operations of propositional logic and certain symbols and operations in set theory, as shown in this table:

Logic	**Set Theory**
$\mathbb{T}$	U
$\mathbb{F}$	$\emptyset$
$p \wedge q$	$A \cap B$
$p \vee q$	$A \cup B$
$\neg p$	$\overline{A}$

Any valid logical formula or computation involving propositional variables and the symbols $\mathbb{T}$, $\mathbb{F}$, $\wedge$, $\vee$, and $\neg$ can be transformed into a valid formula or computation in set theory by replacing the propositions in the formula with subsets of U and replacing the logical symbols with U, $\emptyset$, $\cap$, $\cup$, and the complement operator.

Just as in logic, the operations of set theory can be combined to form complex expressions such as $(A \cup C) \cap \overline{(B \cup \overline{C} \cup D)}$. Parentheses can always be used in such expressions to specify the order in which the operations are to be performed. In the absence of parentheses, we need precedence rules to determine the order of operation. The precedence rules for the Boolean algebra of sets are carried over directly from the Boolean algebra of propositions. When union and intersection are used together without parentheses, intersection has precedence over union. Furthermore, when several operators of the same type are used without parentheses, then they are evaluated in order from left to right. (Of course, since $\cup$ and $\cap$ are both associative operations, it really doesn't matter whether the order of evaluation is left-to-right or right-to-left.) For example, $A \cup B \cap C \cup D$ is evaluated as $(A \cup ((B \cap C)) \cup D$. The complement operation is a special case. Since it is denoted by drawing a line over its operand, there is never any ambiguity about which part of a formula it applies to.

The laws of set theory can be used to simplify complex expressions involving sets. (As usual, of course, the meaning of "simplification" is partly in the eye of the beholder.) For example, for any sets X and Y,

$$\begin{aligned} (X \cup Y) \cap (Y \cup X) &= (X \cup Y) \cap (X \cup Y) && \text{(Commutative Law)} \\ &= (X \cup Y) && \text{(Idempotent Law)} \end{aligned}$$

where in the second step, the Idempotent Law, which says that $A \cap A = A$, is applied with $A = X \cup Y$. For expressions that use the complement operation, it is usually considered to be simpler to apply the operation to an individual set, as in $\overline{A}$, rather than to a formula, as in $\overline{A \cap B}$. DeMorgan's Laws can always be used to simplify an expression in which the complement operation is applied to a formula. For example,

$$
\begin{aligned}
A \cap \overline{B \cup \overline{A}} &= A \cap (\overline{B} \cap \overline{\overline{A}}) && \text{(DeMorgan's Law)} \\
&= A \cap (\overline{B} \cap A) && \text{(Double Complement)} \\
&= A \cap (A \cap \overline{B}) && \text{(Commutative Law)} \\
&= (A \cap A) \cap \overline{B}) && \text{(Associative Law)} \\
&= A \cap \overline{B} && \text{(Idempotent Law)}
\end{aligned}
$$

As a final example of the relationship between set theory and logic, consider the set-theoretical expression $A \cap (A \cup B)$ and the corresponding compound proposition $p \wedge (p \vee q)$. (These correspond since for any x, $x \in A \cap (A \cup B) \equiv (x \in A) \wedge ((x \in A) \vee (x \in B))$.) You might find it intuitively clear that $A \cap (A \cup B) = A$. Formally, this follows from the fact that $p \wedge (p \vee q) \equiv p$, which might be less intuitively clear and is surprising difficult to prove algebraically from the laws of logic. However, there is another way to check that a logical equivalence is valid: Make a truth table. Consider a truth table for $p \wedge (p \vee q)$:

p	q	$p \vee q$	$p \wedge (p \vee q)$
false	false	false	false
false	true	true	false
true	false	true	true
true	true	true	true

The fact that the first column and the last column of this table are identical shows that $p \wedge (p \vee q) \equiv p$. Taking p to be the proposition $x \in A$ and q to be the proposition $x \in B$, it follows that the sets A and $A \cap (A \cup B)$ have the same members and therefore are equal.

Exercises

1. Use the laws of logic to verify the associative laws for union and intersection. That is, show that if A, B, and C are sets, then $A \cup (B \cup C) = (A \cup B) \cup C$ and $A \cap (B \cap C) = (A \cap B) \cap C$.

2. Show that for any sets A and B, $A \subseteq A \cup B$ and $A \cap B \subseteq A$.

3. Recall that the symbol $\oplus$ denotes the logical exclusive or operation. If A and B sets, define the set $A \bigtriangleup B$ by $A \bigtriangleup B = \{x \mid (x \in A) \oplus (x \in B)\}$. Show that $A \bigtriangleup B = (A \smallsetminus B) \cup (B \smallsetminus A)$. ($A \bigtriangleup B$ is known as the ***symmetric difference*** of A and B.)

4. Let A be a subset of some given universal set U. Verify that $\overline{\overline{A}} = A$ and that $A \cup \overline{A} = U$.

5. Verify the second of DeMorgan's Laws for sets, $\overline{A \cap B} = \overline{A} \cup \overline{B}$. For each step in your verification, state why that step is valid.

6. The subset operator, $\subseteq$, is defined in terms of the logical implication operator, $\rightarrow$. However, $\subseteq$ differs from the $\cap$ and $\cup$ operators in that $A \cap B$ and $A \cup B$ are *sets*, while $A \subseteq B$ is a *statement*. So the relationship between $\subseteq$ and $\rightarrow$ isn't quite the same as the relationship between $\cup$ and $\vee$ or between $\cap$ and $\wedge$. Nevertheless, $\subseteq$ and $\rightarrow$ do share some similar properties. This problem shows one example.

  a) Show that the following three compound propositions are logically equivalent: $p \rightarrow q$, $(p \wedge q) \leftrightarrow p$, and $(p \vee q) \leftrightarrow q$.

  b) Show that for any sets A and B, the following three statements are equivalent: $A \subseteq B$, $A \cap B = A$, and $A \cup B = B$.

7. DeMorgan's Laws apply to subsets of some given universal set U. Show that for a subset X of U, $\overline{X} = U \smallsetminus X$. It follows that DeMorgan's Laws can be written as $U \smallsetminus (A \cup B) = (U \smallsetminus A) \cap (U \smallsetminus B)$ and $U \smallsetminus (A \cap B) = (U \smallsetminus A) \cup (U \smallsetminus B)$. Show that these laws hold whether or not A and B are subsets of U. That is, show that for any sets A, B, and C, $C \smallsetminus (A \cup B) = (C \smallsetminus A) \cap (C \smallsetminus B)$ and $C \smallsetminus (A \cap B) = (C \smallsetminus A) \cup (C \smallsetminus B)$.

8. Show that $A \cup (A \cap B) = A$ for any sets A and B.

9. Let X and Y be sets. Simplify each of the following expressions. Justify each step in the simplification with one of the rules of set theory.

  a) $X \cup (Y \cup X)$ **b)** $(X \cap Y) \cap \overline{X}$
  c) $(X \cup Y) \cap \overline{Y}$ **d)** $(X \cup Y) \cup (X \cap Y)$

10. Let A, B, and C be sets. Simplify each of the following expressions. In your answer, the complement operator should only be applied to the individual sets A, B, and C.

  a) $\overline{A \cup B \cup C}$ **b)** $\overline{A \cup B \cap C}$ **c)** $\overline{\overline{A \cup B}}$
  d) $\overline{B \cap \overline{C}}$ **e)** $\overline{A \cap \overline{B \cap \overline{\overline{C}}}}$ **f)** $A \cap \overline{A \cup B}$

11. Use induction to prove the following generalized DeMorgan's Law for set theory: For any natural number $n \geq 2$ and for any sets $X_1, X_2, \ldots, X_n$,

$$\overline{X_1 \cap X_2 \cap \cdots \cap X_n} = \overline{X_1} \cup \overline{X_2} \cup \cdots \cup \overline{X_n}$$

12. State and prove generalized distributive laws for set theory.

2.3 Application: Programming with Sets

On a computer, all data are represented, ultimately, as strings of zeros and ones. At times, computers need to work with sets. How can sets be represented as strings of zeros and ones?

A set is determined by its elements. Given a set A and an entity x, the fundamental question is, does x belong to A or not? If we know the answer to this question for each possible x, then we know the set. For a given x, the answer to the question, "Is x a member of A," is either *yes* or *no*. The answer can be encoded by letting 1 stand for yes and 0 stand for no. The answer, then, is a single ***bit***, that is, a value that can be either zero or one. To represent the set A as a string of zeros and ones, we could use one bit for each possible member of A. If a possible member x is in the set, then the corresponding bit has the value one. If x is not in the set, then the corresponding bit has the value zero.

Now, in cases where the number of possible elements of the set is very large or infinite, it is not practical to represent the set in this way. It would require too many bits, perhaps an infinite number. In such cases, some other representation for the set can be used. However, suppose we are only interested in subsets of some specified small set. Since this set plays the role of a universal set, let's call it U. To represent a subset of U, we just need one bit for each member of U. If the number of members of U is n, then a subset of U is represented by a string of n zeros and ones. Furthermore, every string of n zeros and ones determines a subset of U, namely that subset that contains exactly the elements of U that correspond to ones in the string. A string of n zeros and ones is called an n-bit ***binary number***. So, we see that if U is a set with n elements, then the subsets of U correspond to n-bit binary numbers.

To make things more definite, let U be the set $\{0, 1, 2, \ldots, 31\}$. This set consists of the 32 integers between 0 and 31, inclusive. Then each subset of U can be represented by a 32-bit binary number. We use 32 bits because most computer languages can work directly with 32-bit numbers. For example, the programming languages Java, C, and C++ have a data type named *int*. A value of type *int* is a 32-bit binary number.[1] Before we get a definite correspondence between subsets of U and 32-bit numbers, we have to decide which bit in the number will correspond to each member of U. Following tradition, we assume that the bits are numbered from right to left. That is, the rightmost bit corresponds to the element 0 in U, the second bit from the right corresponds to 1, the third bit from the right to 2, and so on. For example, the 32-bit number

$$10000000000000000000001001110110$$

corresponds to the subset $\{1, 2, 4, 5, 6, 9, 31\}$. Since the leftmost bit of the

[1] Actually, in some versions of C and C++, a value of type *int* is a 16-bit number. A 16-bit number can be used to represent a subset of the set $\{0, 1, 2, \ldots, 15\}$. The principle, of course, is the same.

Hex.	**Binary**	**Hex.**	**Binary**
0	0000_2	8	1000_2
1	0001_2	9	1001_2
2	0010_2	A	1010_2
3	0011_2	B	1011_2
4	0100_2	C	1100_2
5	0101_2	D	1101_2
6	0110_2	E	1110_2
7	0111_2	F	1111_2

Figure 2.3: The 16 hexadecimal digits and the corresponding binary numbers. Each hexadecimal digit corresponds to a 4-bit binary number. Longer binary numbers can be written using two or more hexadecimal digits. For example, $101000011111_2 = 0xA1F$.

number is 1, the number 31 is in the set; since the next bit is 0, the number 30 is not in the set; and so on.

From now on, I will write binary numbers with a subscript of 2 to avoid confusion with ordinary numbers. Furthermore, I will often leave out leading zeros. For example, 1101_2 is the binary number that would be written out in full as

$$00000000000000000000000000001101$$

and which corresponds to the set $\{0, 2, 3\}$. On the other hand 1101 represents the ordinary number one thousand one hundred and one.

Even with this notation, it can be very annoying to write out long binary numbers—and almost impossible to read them. So binary numbers are never written out as sequences of zeros and ones in computer programs. An alternative is to use ***hexadecimal numbers***. Hexadecimal numbers are written using the sixteen symbols 0, 1, 2, 3, 4, 5, 6, 7, 8, 9, A, B, C, D, E, and F. These symbols are knows as the hexadecimal digits. Each hexadecimal digit corresponds to a 4-bit binary number, as shown in Figure 2.3. To represent a longer binary number, several hexadecimal digits can be strung together. For example, the hexadecimal number C7 represents the binary number 11000111_2. In Java and many related languages, a hexadecimal number is written with the prefix "0x". Thus, the hexadecimal number C7 would appear in the program as 0xC7. I will follow the same convention here. Any 32-bit binary number can be written using eight hexadecimal digits (or fewer if leading zeros are omitted). Thus, subsets of $\{0, 1, 2, \ldots, 31\}$ correspond to 8-digit hexadecimal numbers. For example, the subset $\{1, 2, 4, 5, 6, 9, 31\}$ corresponds to 0x80000276, which represents the binary number $10000000000000000000001001110110_2$. Similarly, 0xFF corresponds to $\{0, 1, 2, 3, 4, 5, 6, 7\}$ and 0x1101 corresponds to the binary number 0001000100000001_2 and to the set $\{0, 8, 12\}$.

Now, if you have worked with binary numbers or with hexadecimal numbers, you know that they have another, more common interpretation. They represent ordinary integers. Just as 342 represents the integer $3 \cdot 10^2 + 4 \cdot 10^1 + 2 \cdot 10^0$, the binary number 1101_2 represents the integer $1 \cdot 2^3 + 1 \cdot 2^2 + 0 \cdot 2^1 + 1 \cdot 2^0$, or 13. When used in this way, binary numbers are known as ***base-2 numbers***, just as ordinary numbers are base-10 numbers. Hexadecimal numbers can be interpreted as base-16 numbers. For example, 0x3C7 represents the integer $3 \cdot 16^2 + 12 \cdot 16^1 + 7 \cdot 16^0$, or 874. So, does 1101_2 really represent the integer 13, or does it represent the set $\{0, 2, 3\}$? The answer is that to a person, 1101_2 can represent either. Both are valid interpretations, and the only real question is which interpretation is useful in a given circumstance. On the other hand, to the computer, 1101_2 doesn't represent *anything*. It's just a string of bits, and the computer manipulates the bits according to its program, without regard to their interpretation.

Of course, we still have to answer the question of whether it is ever useful to interpret strings of bits in a computer as representing sets.

If all we could do with sets were to "represent" them, it wouldn't be very useful. We need to be able to compute with sets. That is, we need to be able to perform set operations such as union and complement.

Many programming languages provide operators that perform set operations. In Java and related languages, the operators that perform union, intersection, and complement are written as |, &, and ~. For example, if x and y are 32-bit integers representing two subsets, X and Y, of $\{0, 1, 2, \ldots, 31\}$, then $x \,|\, y$ is a 32-bit integer that represents the set $X \cup Y$. Similarly, $x \,\&\, y$ represents the set $X \cap Y$, and ~x represents the complement, $\overline{X}$.

The operators |, &, and ~ are called ***bitwise logical operators*** because of the way they operate on the individual bits of the numbers to which they are applied. If 0 and 1 are interpreted as the logical values *false* and *true*, then the bitwise logical operators perform the logical operations $\vee$, $\wedge$, and $\neg$ on individual bits. To see why this is true, let's look at the computations that these operators have to perform.

Let k be one of the members of $\{0, 1, 2, \ldots, 31\}$. In the binary numbers x, y, $x \,|\, y$, $x \,\&\, y$, and ~x, the number k corresponds to the bit in position k.

That is, k is in the set represented by a binary number if and only if the bit in position k in that binary number is 1. Considered as sets, $x \,\&\, y$ is the intersection of x and y, so k is a member of the set represented by $x \,\&\, y$ if and only if k is a member of both of the sets represented by x and y. That is, bit k is 1 in the binary number $x \,\&\, y$ if and only if bit k is 1 in x and bit k is 1 in y. When we interpret 1 as *true* and 0 as *false*, we see that bit k of $x \,\&\, y$ is computed by applying the logical "and" operator, $\wedge$, to bit k of x and bit k of y. Similarly, bit k of $x \,|\, y$ is computed by applying the logical "or" operator, $\vee$, to bit k of x and bit k of y. And bit k of ~x is computed by applying the logical "not" operator, $\neg$, to bit k of x. In each case, the logical operator is applied to each bit position separately. (Of course, this discussion is just a translation to the language of bits of the definitions of the set operations in terms of logical operators: $A \cap B = \{x \,|\, x \in A \wedge x \in B\}$, $A \cup B = \{x \,|\, x \in A \vee x \in B\}$, and $\overline{A} = \{x \in U \,|\, \neg(x \in A)\}$.)

For example, consider the binary numbers 1011010_2 and 10111_2, which represent the sets $\{1,3,4,6\}$ and $\{0,1,2,4\}$. Then $1011010_2 \;\&\; 10111_2$ is 10010_2. This binary number represents the set $\{1,4\}$, which is the intersection $\{1,3,4,6\} \cap \{0,1,2,4\}$. It's easier to see what's going on if we write out the computation in columns, the way you probably first learned to do addition:

$$\begin{array}{r} 1\,0\,1\,1\,0\,1\,0 \\ \&\quad 0\,0\,1\,0\,1\,1\,1 \\ \hline 0\,0\,1\,0\,0\,1\,0 \end{array} \qquad \begin{array}{ccccccccc} \{ & 6, & & 4, & 3, & & 1 & & \} \\ \{ & & & 4, & & 2, & 1, & 0 & \} \\ \hline \{ & & & 4, & & & 1 & & \} \end{array}$$

Note that in each column in the binary numbers, the bit in the bottom row is computed as the logical "and" of the two bits that lie above it in the column. I've written out the sets that correspond to the binary numbers to show how the bits in the numbers correspond to the presence or absence of elements in the sets. Similarly, we can see how the union of two sets is computed as a bitwise "or" of the corresponding binary numbers.

$$\begin{array}{r} 1\,0\,1\,1\,0\,1\,0 \\ |\quad 0\,0\,1\,0\,1\,1\,1 \\ \hline 1\,0\,1\,1\,1\,1\,1 \end{array} \qquad \begin{array}{ccccccccc} \{ & 6, & & 4, & 3, & & 1 & & \} \\ \{ & & & 4, & & 2, & 1, & 0 & \} \\ \hline \{ & 6, & & 4, & 3, & 2, & 1, & 0 & \} \end{array}$$

The complement of a set is computed using a bitwise "not" operation. Since we are working with 32-bit binary numbers, the complement is taken with respect to the universal set $\{0,1,2,\dots,31\}$. So, for example,

$$\tilde{}\,1011010_2 = 11111111111111111111111110100101_2$$

Of course, we can apply the operators &, |, and ~ to numbers written in hexadecimal form, or even in ordinary, base-10 form. When doing such calculations by hand, it is probably best to translate the numbers into binary form. For example,

$$\begin{aligned} 0\text{x}AB7 \ \& \ 0\text{x}168E &= 1010\,1011\,0111_2 \ \& \ 1\,0110\,1000\,1110_2 \\ &= 0\,0010\,1000\,0110_2 \\ &= 0\text{x}286 \end{aligned}$$

When computing with sets, it is sometimes necessary to work with individual elements. Typical operations include adding an element to a set, removing an element from a set, and testing whether an element is in a set. However, instead of working with an element itself, it's convenient to work with the set that contains that element as its only member. For example, testing whether $5 \in A$ is the same as testing whether $\{5\} \cap A \neq \emptyset$. The set $\{5\}$ is represented by the binary number 100000_2 or by the hexadecimal number 0x20. Suppose that the set A is represented by the number x. Then, testing whether $5 \in A$ is equivalent to testing whether $0\text{x}20 \ \& \ x \neq 0$. Similarly, the set $A \cup \{5\}$, which is obtained by adding 5 to A, can be computed as $x \mid 0\text{x}20$. The set $A \smallsetminus \{5\}$, which is the set obtained by removing 5 from A if it occurs in A, is represented by $x \ \& \ \tilde{}0\text{x}20$.

The sets $\{0\}$, $\{1\}$, $\{2\}$, $\{3\}$, $\{4\}$, $\{5\}$, $\{6\}$, …, $\{31\}$ are represented by the hexadecimal numbers 0x1, 0x2, 0x4, 0x8, 0x10, 0x20, …, 0x80000000. In typical computer applications, some of these numbers are given names, and these names are thought of as names for the possible elements of a set (although, properly speaking, they are names for sets containing those elements). Suppose, for example, that a, b, c, and d are names for four of the numbers from the above list. Then $a \mid c$ is the set that contains the two elements corresponding to the numbers a and c. If x is a set, then $x \,\&\, \tilde{}d$ is the set obtained by removing d from x. And we can test whether b is in x by testing if $x \,\&\, b \neq 0$.

Here is an actual example, which is used in the Macintosh operating system. Characters can be printed or displayed on the screen in various sizes and styles. A ***font*** is a collection of pictures of characters in a particular size and style. On the Macintosh, a basic font can be modified by specifying any of the following style attributes: *bold*, *italic*, *underline*, *outline*, *shadow*, *condense*, and *extend*. The style of a font is a subset of this set of attributes. A style set can be specified by or-ing together individual attributes. For example, an underlined, bold, italic font has style set *underline* | *bold* | *italic*. For a plain font, with none of the style attributes set, the style set is the empty set, which is represented by the number zero.

The Java programming language uses a similar scheme to specify style attributes for fonts, but currently there are only two basic attributes,

Font.BOLD and Font.ITALIC. A more interesting example in Java is provided by event types. An event in Java represents some kind of user action, such as pressing a key on the keyboard. Events are associated with "components" such as windows, push buttons, and scroll bars. Components can be set to ignore a given type of event. We then say that that event type is disabled for that component. If a component is set to process events of a given type, then that event type is said to be enabled. Each component keeps track of the set of event types that are currently enabled. It will ignore any event whose type is not in that set. Each event type has an associated constant with a name such as AWTEvent.MOUSE_EVENT_MASK. These constants represent the possible elements of a set of event types. A set of event types can be specified by or-ing together a number of such constants. If c is a component and x is a number representing a set of event types, then the command "$c.enableEvents(x)$" enables the events in the set x for the component c. If y represents the set of event types that were already enabled for c, then the effect of this command is to replace y with the union, $y \mid x$. Another command, "$c.disableEvents(x)$", will disable the event types in x for the component c. It does this by replacing the current set, y, with $y \,\&\, \tilde{}x$.

Exercises

1. Suppose that the numbers x and y represent the sets A and B. Show that the set $A \smallsetminus B$ is represented by $x \,\&\, (\tilde{}y)$.

2. Write each of the following binary numbers in hexadecimal:
   a) 10110110_2 **b)** 10_2 **c)** 111100001111_2 **d)** 101001_2

3. Write each of the following hexadecimal numbers in binary:
   a) 0x123 **b)** 0xFADE **c)** 0x137F **d)** 0xFF11

4. Give the value of each of the following expressions as a hexadecimal number:
   a) 0x73 | 0x56A **b)** ~0x3FF0A2FF
   c) (0x44 | 0x95) & 0xE7 **d)** 0x5C35A7 & 0xFF00
   e) 0x5C35A7 & ~0xFF00 **f)** ~(0x1234 & 0x4321)

5. Find a calculator (or a calculator program on a computer) that can work with hexadecimal numbers. Write a short report explaining how to work with hexadecimal numbers on that calculator. You should explain, in particular, how the calculator can be used to do the previous problem.

6. This question assumes that you know how to add binary numbers. Suppose x and y are binary numbers. Under what circumstances will the binary numbers $x + y$ and $x \mid y$ be the same?

7. In addition to hexadecimal numbers, the programming languages Java, C, and C++ support ***octal numbers***. Look up and report on octal numbers in

Java, C, or C++. Explain what octal numbers are, how they are written, and how they are used.

8. In the UNIX (or Linux) operating system, every file has an associated set of permissions, which determine who can use the file and how it can be used. The set of permissions for a given file is represented by a nine-bit binary number. This number is sometimes written as an octal number. Research and report on the UNIX systems of permissions. What set of permissions is represented by the octal number 752? by the octal number 622? Explain what is done by the UNIX commands "chmod g+rw filename" and "chmod o-w filename" in terms of sets. (Hint: Look at the *man* page for the *chmod* command. To see the page, use the UNIX command "man chmod". If you don't know what this means, you probably don't know enough about UNIX to do this exercise.)

9. Java, C, and C++ each have a boolean data type that has the values *true* and *false*. The usual logical and, or, and not operators on boolean values are represented by the operators &&, ||, and !. C and C++ allow integer values to be used in places where boolean values are expected. In this case, the integer zero represents the boolean value *false* while any non-zero integer represents the boolean value *true*. This means that if x and y are integers, then both $x \mathbin{\&} y$ and $x \mathbin{\&\&} y$ are valid expressions, and both can be considered to represent boolean values. Do the expressions $x \mathbin{\&} y$ and $x \mathbin{\&\&} y$ always represent the same boolean value, for any integers x and y? Do the expressions $x \mid y$ and $x \mid\mid y$ always represent the same boolean values? Explain your answers.

10. Suppose that you, as a programmer, want to write a subroutine that will open a window on the computer's screen. The window can have any of the following options: a close box, a zoom box, a resize box, a minimize box, a vertical scroll bar, a horizontal scroll bar. Design a scheme whereby the options for the window can be specified by a single parameter to the subroutine. The parameter should represent a set of options. How would you use your subroutine to open a window that has a close box and both scroll bars and no other options? Inside your subroutine, how would you determine which options have been specified for the window?

2.4 Functions

Both the real world and the world of mathematics are full of what are called, in mathematics, "functional relationships." A functional relationship is a relationship between two sets, which associates exactly one element from the second set to each element of the first set.

For example, each item for sale in a store has a price. The first set in this relationship is the set of items in the store. For each item in the store, there is an associated price, so the second set in the relationship is the set of possible prices. The relationship is a functional relationship because each item has a price. That is, the question "What is the price of this item?" has a single, definite answer for each item in the store.

Similarly, the question "Who is the (biological) mother of this person?" has a single, definite answer for each person. So, the relationship "mother of" defines a functional relationship. In this case, the two sets in the relationship are the same set, namely the set of people.[2] On the other hand, the relationship "child of" is not a functional relationship. The question "Who is the child of this person?" does not have a single, definite answer for each person. A given person might not have any child at all. And a given person might have more than one child. Either of these cases—a person with no child or a person with more than one child—is enough to show that the relationship "child of" is not a functional relationship.

Or consider an ordinary map, such as a map of New York State or a street map of Rome. The whole point of the map, if it is accurate, is that there is a functional relationship between points on the map and points on the surface of the Earth. Perhaps because of this example, a functional relationship is sometimes called a ***mapping***.

There are also many natural examples of functional relationships in mathematics. For example, every rectangle has an associated area. This fact expresses a functional relationship between the set of rectangles and the set of numbers. Every natural number n has a square, n^2. The relationship "square of" is a functional relationship from the set of natural numbers to itself.

In mathematics, of course, we need to work with functional relationships in the abstract. To do this, we introduce the idea of ***function***. You should think of a function as a mathematical object that expresses a functional relationship between two sets. The notation $f \colon A \to B$ expresses the fact that f is a function from the set A to the set B. That is, f is a name for a mathematical object that expresses a functional relationship from the set A to the set B. The notation $f \colon A \to B$ is read as "f is a function from A to B" or more simply as "f maps A to B."

If $f \colon A \to B$ and if $a \in A$, the fact that f is a functional relationship from A to B means that f associates some element of B to a. That element is denoted $f(a)$. That is, for each $a \in A$, $f(a) \in B$ and $f(a)$ is the single, definite answer to the question "What element of B is associated to a by the function f?" The fact that f is a function from A to B means that this question has a single, well-defined answer. Given $a \in A$, $f(a)$ is called the ***value*** of the function f at a.

[2] I'm avoiding here the question of Adam and Eve or of pre-human ape-like ancestors. (Take your pick.)

For example, if I is the set of items for sale in a given store and M is the set of possible prices, then there is function $c\colon I \to M$ which is defined by the fact that for each $x \in I$, $c(x)$ is the price of the item x. Similarly, if P is the set of people, then there is a function $m\colon P \to P$ such that for each person p, $m(p)$ is the mother of p. And if $\mathbb{N}$ is the set of natural numbers, then the formula $s(n) = n^2$ specifies a function $s\colon \mathbb{N} \to \mathbb{N}$. It is in the form of formulas such as $s(n) = n^2$ or $f(x) = x^3 - 3x + 7$ that most people first encounter functions. But you should note that a formula by itself is not a function, although it might well specify a function between two given sets of numbers. Functions are much more general than formulas, and they apply to all kinds of sets, not just to sets of numbers.

Suppose that $f\colon A \to B$ and $g\colon B \to C$ are functions. Given $a \in A$, there is an associated element $f(a) \in B$. Since g is a function from B to C, and since $f(a) \in B$, g associates some element of C to $f(a)$. That element is $g(f(a))$. Starting with an element a of A, we have produced an associated element $g(f(a))$ of C. This means that we have defined a new function from the set A to the set C. This function is called the ***composition*** of g with f, and it is denoted by $g \circ f$. That is, if $f\colon A \to B$ and $g\colon B \to C$ are functions, then $g \circ f\colon A \to C$ is the function which is defined by

$$(g \circ f)(a) = g(f(a))$$

for each $a \in A$. For example, suppose that p is the function that associates to each item in a store the price of the item, and suppose that t is a function that associates the amount of tax on a price to each possible price. The composition, $t \circ p$, is the function that associates to each item the amount of tax on that item. Or suppose that $s\colon \mathbb{N} \to \mathbb{N}$ and $r\colon \mathbb{N} \to \mathbb{N}$ are the functions defined by the formulas $s(n) = n^2$ and $r(n) = 3n + 1$, for each $n \in \mathbb{N}$. Then $r \circ s$ is a function from $\mathbb{N}$ to $\mathbb{N}$, and for $n \in \mathbb{N}$, $(r \circ s)(n) = r(s(n)) = r(n^2) = 3n^2 + 1$. In this case, we also have the function $s \circ r$, which satisfies $(s \circ r)(n) = s(r(n)) = s(3n+1) = (3n+1)^2 = 9n^2 + 6n + 1$. Note in particular that $r \circ s$ and $s \circ r$ are not the same function. The operation $\circ$ is not commutative.

If A is a set and $f\colon A \to A$, then $f \circ f$, the composition of f with itself, is defined. For example, using the function s from the preceding example, $s \circ s$ is the function from $\mathbb{N}$ to $\mathbb{N}$ given by the formula $(s \circ s)(n) = s(s(n)) = s(n^2) = (n^2)^2 = n^4$. If m is the function from the set of people to itself which associates to each person that person's mother, then $m \circ m$ is the function that associates to each person that person's maternal grandmother.

If a and b are entities, then (a, b) denotes the ***ordered pair*** containing a and b. The ordered pair (a, b) differs from the set $\{a, b\}$ because a set is not ordered. That is, $\{a, b\}$ and $\{b, a\}$ denote the same set, but if $a \neq b$, then (a, b) and (b, a) are different ordered pairs. More generally, two ordered pairs (a, b) and (c, d) are equal if and only if both $a = c$ and $b = d$. If (a, b) is an ordered pair, then a and b are referred to as the ***coordinates*** of the ordered pair. In particular, a is the first coordinate and b is the second coordinate.

If A and B are sets, then we can form the set $A \times B$ which is defined by

$$A \times B = \{(a, b) \mid a \in A \text{ and } b \in B\}.$$

This set is called the ***cross product*** or ***Cartesian product*** of the sets A and B. The set $A \times B$ contains every ordered pair whose first coordinate is an element of A and whose second coordinate is an element of B. For example, if $X = \{c, d\}$ and $Y = \{1, 2, 3\}$, then $X \times Y = \{(c, 1), (c, 2), (c, 3), (d, 1), (d, 2), (d, 3)\}$. It is possible to extend this idea to the cross product of more than two sets. The cross product of the three sets A, B, and C is denoted $A \times B \times C$. It consists of all ***ordered triples*** (a, b, c) where $a \in A$, $b \in B$, and $c \in C$. The definition for four or more sets is similar. The general term for a member of a cross product is ***tuple*** or, more specifically, ***ordered n-tuple***. For example, (a, b, c, d, e) is an ordered 5-tuple.

Given a function $f \colon A \to B$, consider the set $\{(a, b) \in A \times B \mid a \in A \text{ and } b = f(a)\}$. This set of ordered pairs consists of all pairs (a, b) such that $a \in A$ and b is the element of B that is associated to a by the function f. The set $\{(a, b) \in A \times B \mid a \in A \text{ and } b = f(a)\}$ is called the ***graph*** of the function f. Since f is a function, each element $a \in A$ occurs once and only once as a first coordinate among the ordered pairs in the graph of f. Given $a \in A$, we can determine $f(a)$ by finding that ordered pair and looking at the second coordinate. In fact, it is convenient to consider the function and its graph to be the same thing, and to use this as our official mathematical definition.[3]

Definition 2.2. Let A and B be sets. A ***function*** from A to B is a subset of $A \times B$ which has the property that for each $a \in A$, the set contains one and only one ordered pair whose first coordinate is a. If (a, b) is that ordered pair, then b is called the value of the function at a and is denoted $f(a)$. If $b = f(a)$, then we also say that the function f ***maps*** a to b. The fact that f is a function from A to B is indicated by the notation $f \colon A \to B$.

[3]This is a convenient definition for the mathematical world, but as is often the case in mathematics, it leaves out an awful lot of the real world. Functional relationships in the real world are *meaningful*, but we model them in mathematics with meaningless sets of ordered pairs. We do this for the usual reason: to have something precise and rigorous enough that we can make logical deductions and prove things about it.

For example, if $X = \{a, b\}$ and $Y = \{1, 2, 3\}$, then the set $\{(a, 2), (b, 1)\}$ is a function from X to Y, and $\{(1, a), (2, a), (3, b)\}$ is a function from Y to X. On the other hand, $\{(1, a), (2, b)\}$ is not a function from Y to X, since it does not specify any value for 3. And $\{(a, 1), (a, 2), (b, 3)\}$ is not a function from X to Y because it specifies two different values, 1 and 2, associated with the same element, a, of X.

Even though the technical definition of a function is a set of ordered pairs, it's usually better to think of a function from A to B as something that associates some element of B to every element of A. The set of ordered pairs is one way of expressing this association. If the association is expressed in some other way, it's easy to write down the set of ordered pairs. For example, the function $s\colon \mathbb{N} \to \mathbb{N}$ which is specified by the formula $s(n) = n^2$ can be written as the set of ordered pairs $\{(n, n^2) \mid n \in \mathbb{N}\}$.

Suppose that $f\colon A \to B$ is a function from the set A to the set B. We say that A is the ***domain*** of the function f and that B is the ***range*** of the function. We define the ***image*** of the function f to be the set $\{b \in B \mid \exists a \in A\,(b = f(a))\}$. Put more simply, the image of f is the set $\{f(a) \mid a \in A\}$. That is, the image is the set of all values, $f(a)$, of the function, for all $a \in A$. (You should note that in some cases—particularly in calculus courses—the term "range" is used to refer to what I am calling the image.) For example, for the function $s\colon \mathbb{N} \to \mathbb{N}$ that is specified by $s(n) = n^2$, both the domain and the range are $\mathbb{N}$, and the image is the set $\{n^2 \mid n \in \mathbb{N}\}$, or $\{0, 1, 4, 9, 16, \dots\}$.

Note that the image of a function is a subset of its range. It can be a proper subset, as in the above example, but it is also possible for the image of a function to be equal to the range. In that case, the function is said to be ***onto***. Sometimes, the fancier term ***surjective*** is used instead. Formally, a function $f\colon A \to B$ is said to be onto (or surjective) if every element of B is equal to $f(a)$ for some element of A. In terms of logic, f is onto if and only if

$$\forall b \in B\,\big(\exists a \in A\,(b = f(a))\big).$$

For example, let $X = \{a, b\}$ and $Y = \{1, 2, 3\}$, and consider the function from Y to X specified by the set of ordered pairs $\{(1, a), (2, a), (3, b)\}$. This function is onto because its image, $\{a, b\}$, is equal to the range, X. However, the function from X to Y given by $\{(a, 1), (b, 3)\}$ is not onto, because its image, $\{1, 3\}$, is a proper subset of its range, Y. As a further example, consider the function f from $\mathbb{Z}$ to $\mathbb{Z}$ given by $f(n) = n - 52$. To show that f is onto, we need to pick an arbitrary b in the range $\mathbb{Z}$ and show that there is some number a in the domain $\mathbb{Z}$ such that $f(a) = b$. So let b be an arbitrary integer; we want to find an a such that $a - 52 = b$. Clearly this equation will be true when $a = b + 52$. So every element b is the image of the number $a = b + 52$, and f is therefore onto. Note that if f had been specified to have domain $\mathbb{N}$, then f would *not* be onto, as for some $b \in \mathbb{Z}$ the number $a = b + 52$ is not in the domain $\mathbb{N}$ (for example, the integer -73 is not in the image of f, since -21 is not in $\mathbb{N}$.)

If $f\colon A \to B$ and if $a \in A$, then a is associated to only one element of B. This is part of the definition of a function. However, no such restriction holds for elements of B. If $b \in B$, it is possible for b to be associated to zero, one, two, three, ..., or even to an infinite number of elements of A. In the case where each element of the range is associated to at most one element of the domain, the function is said to be ***one-to-one***. Sometimes, the term ***injective*** is used instead. The function f is one-to-one (or injective) if for any two distinct elements x and y in the domain of f, $f(x)$ and $f(y)$ are also distinct. In terms of logic, $f\colon A \to B$ is one-to-one if and only if

$$\forall x \in A\ \forall y \in A\,\big(x \neq y \to f(x) \neq f(y)\big).$$

Since a proposition is equivalent to its contrapositive, we can write this condition equivalently as

$$\forall x \in A\ \forall y \in A\,\big(f(x) = f(y) \to x = y\big).$$

Sometimes, it is easier to work with the definition of one-to-one when it is expressed in this form. The function that associates every person to his or her mother is not one-to-one because it is possible for two different people to have the same mother. The function $s\colon \mathbb{N} \to \mathbb{N}$ specified by $s(n) = n^2$ is one-to-one. However, we can define a function $r\colon \mathbb{Z} \to \mathbb{Z}$ by the same formula: $r(n) = n^2$, for $n \in \mathbb{Z}$. The function r is *not* one-to-one since two different integers can have the same square. For example, $r(-2) = r(2)$.

A function that is both one-to-one and onto is said to be ***bijective***. The function that associates each point in a map of New York State to a point in the state itself is presumably bijective. For each point on the map, there is a corresponding point in the state, and *vice versa*. If we specify the function f from the set $\{1, 2, 3\}$ to the set $\{a, b, c\}$ as the set of ordered pairs $\{(1, b), (2, a), (3, c)\}$, then f is a bijective function. Or consider the function from $\mathbb{Z}$ to $\mathbb{Z}$ given by $f(n) = n - 52$. We have already shown that f is onto. We can show that it is also one-to-one: pick an arbitrary x and y in $\mathbb{Z}$ and assume that $f(x) = f(y)$. This means that $x - 52 = y - 52$, and adding 52 to both sides of the equation gives $x = y$. Since x and y were arbitrary, we have proved $\forall x \in \mathbb{Z}\ \forall y \in \mathbb{Z}\,(f(x) = f(y) \to x = y)$, that is, that f is one-to-one. Altogether, then, f is a bijection.

One difficulty that people sometimes have with mathematics is its generality. A set is a collection of entities, but an "entity" can be anything at all, including other sets. Once we have defined ordered pairs, we can use ordered pairs as elements of sets. We could also make ordered pairs of sets. Now that we have defined functions, every function is itself an entity. This means that we can have sets that contain functions. We can even have a function whose domain and range are sets of functions. Similarly, the domain or range of a function might be a set of sets, or a set of ordered pairs. Computer scientists have a good name for this. They would say that sets, ordered pairs, and functions are ***first-class objects***. Once a set, ordered pair, or function has been defined, it can be used just like any other entity. If they were not first-class objects, there could be restrictions on the way they can be used. For example, it might not be possible to use functions as members of sets. (This would make them "second class.")

For example, suppose that A, B, and C are sets. Then since $A \times B$ is a set, we might have a function $f\colon A \times B \to C$. If $(a, b) \in A \times B$, then the value of f at (a, b) would be denoted $f((a, b))$. In practice, though, one set of parentheses is usually dropped, and the value of f at (a, b) is denoted $f(a, b)$. As a particular example, we might define a function $p\colon \mathbb{N} \times \mathbb{N} \to \mathbb{N}$ with the formula $p(n, m) = nm + 1$. Similarly, we might define a function $q\colon \mathbb{N} \times \mathbb{N} \times \mathbb{N} \to \mathbb{N} \times \mathbb{N}$ by $q(n, m, k) = (nm - k, nk - n)$.

Suppose that A and B are sets. There are, in general, many functions that map A to B. We can gather all those functions into a set. This set, whose elements are all the functions from A to B, is denoted B^A. (We'll see later why this notation is reasonable.) Using this notation, saying $f\colon A \to B$ is exactly the same as saying $f \in B^A$. Both of these notations assert that f is a function from A to B. Of course, we can also form an unlimited number of other sets, such as the power set $\mathcal{P}(B^A)$, the cross product $B^A \times A$, or the set $A^{A \times A}$, which contains all the functions from the set $A \times A$ to the set A. And of course, any of these sets can be the domain or range of a function. An example of this is the function $\mathcal{E}\colon B^A \times A \to B$ defined by the formula $\mathcal{E}(f, a) = f(a)$. Let's see if we can make sense of this notation. Since the domain of $\mathcal{E}$ is $B^A \times A$, an element in the domain is an ordered pair in which the first coordinate is a function from A to B and the second coordinate is an element of A. Thus, $\mathcal{E}(f, a)$ is defined for a function $f\colon A \to B$ and an element $a \in A$. Given such an f and a, the notation $f(a)$ specifies an element of B, so the definition of $\mathcal{E}(f, a)$ as $f(a)$ makes sense. The function $\mathcal{E}$ is called the "evaluation function" since it captures the idea of evaluating a function at an element of its domain.

Exercises

1. Let $A = \{1, 2, 3, 4\}$ and let $B = \{a, b, c\}$. Find the sets $A \times B$ and $B \times A$.

2. Let A be the set $\{a, b, c, d\}$. Let f be the function from A to A given by the set of ordered pairs $\{(a, b), (b, b), (c, a), (d, c)\}$, and let g be the function given by the set of ordered pairs $\{(a, b), (b, c), (c, d), (d, d)\}$. Find the set of ordered pairs for the composition $g \circ f$.

3. Let $A = \{a, b, c\}$ and let $B = \{0, 1\}$. Find all possible functions from A to B. Give each function as a set of ordered pairs. (Hint: Every such function corresponds to one of the subsets of A.)

4. Consider the functions from $\mathbb{Z}$ to $\mathbb{Z}$ which are defined by the following formulas. Decide whether each function is onto and whether it is one-to-one; prove your answers.

a) $f(n) = 2n$ **b)** $g(n) = n + 1$ **c)** $h(n) = n^2 + n + 1$

d) $s(n) = \begin{cases} \text{n/2}, & \text{if } n \text{ is even} \\ \text{(n+1)/2}, & \text{if } n \text{ is odd} \end{cases}$

5. Prove that composition of functions is an associative operation. That is, prove that for functions $f : A \to B$, $g : B \to C$, and $h : C \to D$, the compositions $(h \circ g) \circ f$ and $h \circ (g \circ f)$ are equal.

6. Suppose that $f : A \to B$ and $g : B \to C$ are functions and that $g \circ f$ is one-to-one.

a) Prove that f is one-to-one. (Hint: use a proof by contradiction.)

b) Find a specific example that shows that g is not necessarily one-to-one.

7. Suppose that $f : A \to B$ and $g : B \to C$, and suppose that the composition $g \circ f$ is an onto function.

a) Prove that g is an onto function.

b) Find a specific example that shows that f is not necessarily onto.

2.5 Application: Programming with Functions

Functions are fundamental in computer programming, although not everything in programming that goes by the name of "function" is a function according to the mathematical definition.

In computer programming, a function is a routine that is given some data as input and that will calculate and return an answer based on that data. For example, in the C++ programming language, a function that calculates the square of an integer could be written

```
int square(int n) {
   return n*n;
}
```

In C++, *int* is a data type. From the mathematical point of view, a data type is a set. The data type *int* is the set of all integers that can be represented as 32-bit binary numbers. Mathematically, then, $int \subseteq \mathbb{Z}$.

(You should get used to the fact that sets and functions can have names that consist of more than one character, since it's done all the time in computer programming.) The first line of the above function definition, "`int square(int n)`", says that we are defining a function named square whose range is *int* and whose domain is *int*. In the usual notation for functions, we would express this as $square\colon int \to int$, or possibly as $square \in int^{int}$, where int^{int} is the set of all functions that map the set *int* to the set *int*.

The first line of the function, `int square(int n)`, is called the ***prototype*** of the function. The prototype specifies the name, the domain, and the range of the function and so carries exactly the same information as the notation "$f\colon A \to B$". The "n" in "`int square(int n)`" is a name for an arbitrary element of the data type *int*. In computer jargon, n is called a ***parameter*** of the function. The rest of the definition of *square* tells the computer to calculate the value of $square(n)$ for any $n \in int$ by multiplying n times n. The statement "`return n*n`" says that $n * n$ is the value that is computed, or "returned," by the function. (The $*$ stands for multiplication.)

C++ has many data types in addition to *int*. There is a boolean data type named *bool*. The values of type *bool* are *true* and *false*. Mathematically, *bool* is a name for the set $\{true, false\}$. The type *float* consists of real numbers, which can include a decimal point. Of course, on a computer, it's not possible to represent the entire infinite set of real numbers, so *float* represents some subset of the mathematical set of real numbers. There is also a data type whose values are strings of characters, such as "Hello world" or "xyz152QQZ". The name for this data type in C++ is *string*. All these types, and many others, can be used in functions. For example, in C++, $m \,\%\, n$ is the remainder when the integer m is divided by the integer n. We can define a function to test whether an integer is even as follows:

```
bool even(int k) {
   if ( k % 2 == 1 )
      return false;
   else
      return true;
}
```

You don't need to worry about all the details here, but you should understand that the prototype, `bool even(int k)`, says that *even* is a function from the set *int* to the set *bool*. That is, $even\colon int \to bool$. Given an integer N, $even(N)$ has the value *true* if N is an even integer, and it has the value *false* if N is an odd integer.

A function can have more than one parameter. For example, we might define a function with prototype `int index(string str, string sub)`. If s and t are strings, then $index(s,t)$ would be the *int* that is the value of the function at the ordered pair (s,t). We see that the domain of index is the cross product *string* × *string*, and we can write $index\colon string \times string \rightarrow int$ or, equivalently, $index \in int^{string \times string}$.

Not every C++ function is actually a function in the mathematical sense. In mathematics, a function must associate a single value in its range to each value in its domain. There are two things that can go wrong: The value of the function might not be defined for every element of the domain, and the function might associate several different values to the same element of the domain. Both of these things can happen with C++ functions.

In computer programming, it is very common for a "function" to be undefined for some values of its parameter. In mathematics, a ***partial function*** from a set A to a set B is defined to be a function from a subset of A to B. A partial function from A to B can be undefined for some elements of A, but when it is defined for some $a \in A$, it associates just one element of B to a. Many functions in computer programs are actually partial functions. (When dealing with partial functions, an ordinary function, which is defined for every element of its domain, is sometimes referred to as a ***total function***. Note that—with the mind-boggling logic that is typical of mathematicians—a total function is a type of partial function, because a set is a subset of itself.)

It's also very common for a "function" in a computer program to produce a variety of values for the same value of its parameter. A common example is a function with prototype `int random(int N)`, which returns a random integer between 1 and N. The value of $random(5)$ could be 1, 2, 3, 4, or 5. This is not the behavior of a mathematical function!

Even though many functions in computer programs are not really mathematical functions, I will continue to refer to them as functions in this section. Mathematicians will just have to stretch their definitions a bit to accommodate the realities of computer programming.

In most programming languages, functions are not first-class objects. That is, a function cannot be treated as a data value in the same way as a *string* or an *int*. However, C++ does take a step in this direction. It is possible for a function to be a parameter to another function. For example, consider the function prototype

```
float sumten( float f(int) )
```

This is a prototype for a function named *sumten* whose parameter is a function. The parameter is specified by the prototype "`float f(int)`".

This means that the parameter must be a function from *int* to *float*. The parameter name, f, stands for an arbitrary such function. Mathematically, $f \in \mathit{float}^{\mathit{int}}$, and so *sumten*: $\mathit{float}^{\mathit{int}} \to \mathit{float}$.

My idea is that $\mathit{sumten}(f)$ would compute $f(1) + f(2) + \cdots + f(10)$. A more useful function would be able to compute $f(a) + f(a+1) + \cdots + f(b)$ for any integers a and b. This just means that a and b should be parameters to the function. The prototype for the improved function would look like

```
float sum( float f(int), int a, int b )
```

The parameters to *sum* form an ordered triple in which the first coordinate is a function and the second and third coordinates are integers. So, we could write

$$\mathit{sum}\colon \mathit{float}^{\mathit{int}} \times \mathit{int} \times \mathit{int} \to \mathit{float}$$

It's interesting that computer programmers deal routinely with such complex objects.

One thing you can't do in C++ is write a function that creates new functions from scratch. The only functions that exist are those that are coded into the source code of the program. There are programming languages that do allow new functions to be created from scratch while a program is running. In such languages, functions are first-class objects. These languages support what is called ***functional programming***.

One of the most accessible languages that supports functional programming is JavaScript, a language that is used on Web pages. (Although the names are similar, JavaScript and Java are only distantly related.) In JavaScript, the function that computes the square of its parameter could be defined as

```
function square(n) {
   return n*n;
}
```

This is similar to the C++ definition of the same function, but you'll notice that no type is specified for the parameter n or for the value computed by the function. Given this definition of square, $\mathit{square}(x)$ would be legal for any x of any type. (Of course, the value of $\mathit{square}(x)$ would be undefined for most types, so *square* is a *very* partial function, like most functions in JavaScript.) In effect, all possible data values in JavaScript are bundled together into one set, which I will call *data*. We then have *square*: $\mathit{data} \to \mathit{data}$.[4]

[4]Not all functional programming languages lump data types together in this way. There is a functional programming language named Haskell, for example, that is as strict about types as C++. For information about Haskell, see http://www.Haskell.org/.

In JavaScript, a function really is a first-class object. We can begin to see this by looking at an alternative definition of the function *square*:

```
square = function(n) { return n*n; }
```

Here, the notation "`function(n) { return n*n; }`" creates a function that computes the square of its parameter, but it doesn't give any name to this function. This function object is then assigned to a variable named *square*. The value of *square* can be changed later, with another assignment statement, to a different function or even to a different type of value. This notation for creating function objects can be used in other places besides assignment statements. Suppose, for example, that a function with prototype `function sum(f,a,b)` has been defined in a JavaScript program to compute $f(a)+f(a+1)+\cdots+f(b)$. Then we could compute $1^2+2^2+\cdots+100^2$ by saying

```
sum( function(n) { return n*n; }, 1, 100 )
```

Here, the first parameter is the function that computes squares. We have created and used this function without ever giving it a name.

It is even possible in JavaScript for a function to return another function as its value. For example,

```
function monomial(a, n) {
   return ( function(x) { a*Math.pow(x,n); } );
}
```

Here, `Math.pow(x,n)` computes x^n, so for any numbers a and n, the value of *monomial*(a,n) is a function that computes ax^n. Thus,

```
f = monomial(2,3);
```

would define f to be the function that satisfies $f(x) = 2x^3$, and if *sum* is the function described above, then

```
sum( monomial(8,4), 3, 6 )
```

would compute $8*3^4 + 8*4^4 + 8*5^4 + 8*6^4$. In fact, *monomial* can be used to create an unlimited number of new functions from scratch. It is even possible to write *monomial*(2,3)(5) to indicate the result of applying the function *monomial*(2,3) to the value 5. The value represented by *monomial*(2,3)(5) is $2*5^3$, or 250. This is real functional programming and might give you some idea of its power.

Exercises

1. For each of the following C++ function prototypes, translate the prototype into a standard mathematical function specification, such as *func*: *float* → *int*.
   a) `int strlen(string s)`
   b) `float pythag(float x, float y)`
   c) `int round(float x)`
   d) `string sub(string s, int n, int m)`
   e) `string unlikely( int f(string) )`
   f) `int h( int f(int), int g(int) )`
2. Write a C++ function prototype for a function that belongs to each of the following sets.
   a) $string^{string}$
   b) $bool^{float \times float}$
   c) $float^{int^{int}}$
3. It is possible to define new types in C++. For example, the definition

```
struct point {
   float x;
   float y;
}
```

   defines a new type named *point*. A value of type *point* contains two values of type *float*. What mathematical operation corresponds to the construction of this data type? Why?
4. Let *square*, *sum* and *monomial* be the JavaScript functions described in this section. What is the value of each of the following?
   a) *sum*(*square*, 2, 4)
   b) *sum*(*monomial*(5,2), 1, 3)
   c) *monomial*(*square*(2), 7)
   d) *sum*(function(n) { return $2 * n$; }, 1, 5)
   e) *square*(*sum*(*monomial*(2,3), 1, 2))
5. Write a JavaScript function named *compose* that computes the composition of two functions. That is, *compose*(f,g) is $f \circ g$, where f and g are functions of one parameter. Recall that $f \circ g$ is the function defined by $(f \circ g)(x) = f(g(x))$.

2.6 Counting Past Infinity

As children, we all learned to answer the question "How many?" by counting with numbers: 1, 2, 3, 4, ... But the question of "How many?" was asked and answered long before the abstract concept of number was invented. The answer can be given in terms of "as many as." How many cousins do you have? As many cousins as I have fingers on both hands. How many sheep do you own? As many sheep as there are notches on this stick. How many baskets of wheat must I pay in taxes? As many baskets as there are stones in this box. The question of how many things are in one collection of objects is answered by exhibiting another, more convenient, collection of objects that has just as many members.

In set theory, the idea of one set having just as many members as another set is expressed in terms of ***one-to-one correspondence***. A one-to-one correspondence between two sets A and B pairs each element of A with an element of B in such a way that every element of B is paired with one and only one element of A. The process of counting, as it is learned by children, establishes a one-to-one correspondence between a set of n objects and the set of numbers from 1 to n. The rules of counting are the rules of one-to-one correspondence: Make sure you count every object, make sure you don't count the same object more than once. That is, make sure that each object corresponds to *one* and *only one* number. Earlier in this chapter, we used the fancy name "bijective function" to refer to this idea, but we can now see it as as an old, intuitive way of answering the question "How many?"

In counting, as it is learned in childhood, the set $\{1, 2, 3, \ldots, n\}$ is used as a typical set that contains n elements. In mathematics and computer science, it has become more common to start counting with zero instead of with one, so we define the following sets to use as our basis for counting:

$$
\begin{array}{ll}
N_0 = \emptyset, & \text{a set with 0 elements} \\
N_1 = \{0\}, & \text{a set with 1 element} \\
N_2 = \{0,1\}, & \text{a set with 2 elements} \\
N_3 = \{0,1,2\}, & \text{a set with 3 elements} \\
N_4 = \{0,1,2,3\}, & \text{a set with 4 elements}
\end{array}
$$

and so on. In general, $N_n = \{0, 1, 2, \ldots, n-1\}$ for each $n \in \mathbb{N}$. For each natural number n, N_n is a set with n elements. Note that if $n \neq m$, then there is no one-to-one correspondence between N_n and N_m. This is obvious, but like many obvious things is not all that easy to prove rigorously, and we omit the argument here.

Theorem 2.6. *For each $n \in \mathbb{N}$, let N_n be the set $N_n = \{0, 1, \ldots, n-1\}$. If $n \neq m$, then there is no bijective function from N_m to N_n.*

We can now make the following definitions:

Definition 2.3. A set A is said to be ***finite*** if there is a one-to-one correspondence between A and N_n for some natural number n. We then say that n is the ***cardinality*** of A. The notation $|A|$ is used to indicate the cardinality of A. That is, if A is a finite set, then $|A|$ is the natural number n such that there is a one-to-one correspondence between A and N_n. A set that is not finite is said to be ***infinite***. That is, a set B is infinite if for every $n \in \mathbb{N}$, there is *no* one-to-one correspondence between B and N_n.

Fortunately, we don't always have to count every element in a set individually to determine its cardinality. Consider, for example, the set $A \times B$, where A and B are finite sets. If we already know $|A|$ and $|B|$, then we can determine $|A \times B|$ by computation, without explicit counting of elements. In fact, $|A \times B| = |A| \cdot |B|$. The cardinality of the cross product $A \times B$ can be computed by multiplying the cardinality of A by the cardinality of B. To see why this is true, think of how you might count the elements of $A \times B$. You could put the elements into piles, where all the ordered pairs in a pile have the same first coordinate. There are as many piles as there are elements of A, and each pile contains as many ordered pairs as there are elements of B. That is, there are $|A|$ piles, with $|B|$ items in each. By the definition of multiplication, the total number of items in all the piles is $|A| \cdot |B|$. A similar result holds for the cross product of more that two finite sets. For example, $|A \times B \times C| = |A| \cdot |B| \cdot |C|$.

It's also easy to compute $|A \cup B|$ in the case where A and B are disjoint finite sets. (Recall that two sets A and B are said to be disjoint if they have no members in common, that is, if $A \cap B = \emptyset$.) Suppose $|A| = n$ and $|B| = m$. If we wanted to count the elements of $A \cup B$, we could use the n numbers from 0 to $n - 1$ to count the elements of A and then use the m numbers from n to $n + m - 1$ to count the elements of B. This amounts to a one-to-one correspondence between $A \cup B$ and the set N_{n+m}. We see that $|A \cup B| = n + m$. That is, for disjoint finite sets A and B, $|A \cup B| = |A| + |B|$.

What about $A \cup B$, where A and B are not disjoint? We have to be careful not to count the elements of $A \cap B$ twice. After counting the elements of A, there are only $|B| - |A \cap B|$ new elements in B that still need to be counted. So we see that for any two finite sets A and B, $|A \cup B| = |A| + |B| - |A \cap B|$.

What about the number of subsets of a finite set A? What is the relationship between $|A|$ and $|\mathcal{P}(A)|$? The answer is provided by the following theorem.

Theorem 2.7. *A finite set with cardinality n has 2^n subsets.*

Proof. Let $P(n)$ be the statement "Any set with cardinality n has 2^n subsets." We will use induction to show that $P(n)$ is true for all $n \in N$.

Base case: For $n = 0$, $P(n)$ is the statement that a set with cardinality

0 has 2^0 subsets. The only set with 0 elements is the empty set. The empty set has exactly 1 subset, namely itself. Since $2^0 = 1$, $P(0)$ is true.

Inductive case: Let k be an arbitrary element of $\mathbb{N}$, and assume that $P(k)$ is true. That is, assume that any set with cardinality k has 2^k elements. (This is the induction hypothesis.) We must show that $P(k+1)$ follows from this assumption. That is, using the assumption that any set with cardinality k has 2^k subsets, we must show that any set with cardinality $k+1$ has 2^{k+1} subsets.

Let A be an arbitrary set with cardinality $k+1$. We must show that $|\mathcal{P}(A)| = 2^{k+1}$. Since $|A| > 0$, A contains at least one element. Let x be some element of A, and let $B = A \smallsetminus \{x\}$. The cardinality of B is k, so we have by the induction hypothesis that $|\mathcal{P}(B)| = 2^k$. Now, we can divide the subsets of A into two classes: subsets of A that do not contain x and subsets of A that do contain x. Let Y be the collection of subsets of A that do not contain x, and let X be the collection of subsets of A that do contain x. X and Y are disjoint, since it is impossible for a given subset of A both to contain and to not contain x. It follows that $|\mathcal{P}(A)| = |X \cup Y| = |X| + |Y|$.

Now, a member of Y is a subset of A that does not contain x. But that is exactly the same as saying that a member of Y is a subset of B. So $Y = \mathcal{P}(B)$, which we know contains 2^k members. As for X, there is a one-to-one correspondence between $\mathcal{P}(B)$ and X. Namely, the function $f : \mathcal{P}(B) \to X$ defined by $f(C) = C \cup \{x\}$ is a bijective function. (The proof of this is left as an exercise.) From this, it follows that $|X| = |\mathcal{P}(B)| = 2^k$. Putting these facts together, we see that $|\mathcal{P}(A)| = |X| + |Y| = 2^k + 2^k = 2 \cdot 2^k = 2^{k+1}$. This completes the proof that $P(k) \to P(k+1)$. □

We have seen that the notation A^B represents the set of all functions from B to A. Suppose A and B are finite, and that $|A| = n$ and $|B| = m$. Then $\left|A^B\right| = n^m = |A|^{|B|}$. (This fact is one of the reasons why the notation A^B is reasonable.) One way to see this is to note that there is a one-to-one correspondence between A^B and a cross product $A \times A \times \cdots A$, where the number of terms in the cross product is m. (This will be shown in one of the exercises at the end of this section.) It follows that $\left|A^B\right| = |A| \cdot |A| \cdots |A| = n \cdot n \cdots n$, where the factor n occurs m times in the product. This product is, by definition, n^m.

This discussion about computing cardinalities is summarized in the following theorem:

Theorem 2.8. *Let A and B be finite sets. Then*

- $|A \times B| = |A| \cdot |B|$.
- $|A \cup B| = |A| + |B| - |A \cap B|$.

- *If A and B are disjoint then* $|A \cup B| = |A| + |B|$.
- $|A^B| = |A|^{|B|}$.
- $|\mathcal{P}(A)| = 2^{|A|}$.

When it comes to counting and computing cardinalities, this theorem is only the beginning of the story. There is an entire large and deep branch of mathematics known as ***combinatorics*** that is devoted mostly to the problem of counting. But the theorem is already enough to answer many questions about cardinalities.

For example, suppose that $|A| = n$ and $|B| = m$. We can form the set $\mathcal{P}(A \times B)$, which consists of all subsets of $A \times B$. Using the theorem, we can compute that $|\mathcal{P}(A \times B)| = 2^{|A \times B|} = 2^{|A| \cdot |B|} = 2^{nm}$. If we assume that A and B are disjoint, then we can compute that $\left|A^{A \cup B}\right| = |A|^{|A \cup B|} = n^{n+m}$.

To be more concrete, let $X = \{a, b, c, d, e\}$ and let $Y = \{c, d, e, f\}$ where a, b, c, d, e, and f are distinct. Then $|X \times Y| = 5 \cdot 4 = 20$ while $|X \cup Y| = 5 + 4 - |\{c, d, e\}| = 6$ and $\left|X^Y\right| = 5^4 = 625$.

We can also answer some simple practical questions. Suppose that in a restaurant you can choose one appetizer and one main course. What is the number of possible meals? If A is the set of possible appetizers and C is the set of possible main courses, then your meal is an ordered pair belonging to the set $A \times C$. The number of possible meals is $|A \times C|$, which is the product of the number of appetizers and the number of main courses.

Or suppose that four different prizes are to be awarded, and that the set of people who are eligible for the prizes is A. Suppose that $|A| = n$. How many different ways are there to award the prizes? One way to answer this question is to view a way of awarding the prizes as a function from the set of prizes to the set of people. Then, if P is the set of prizes, the number of different ways of awarding the prizes is $\left|A^P\right|$. Since $|P| = 4$ and $|A| = n$, this is n^4. Another way to look at it is to note that the people who win the prizes form an ordered tuple (a, b, c, d), which is an element of $A \times A \times A \times A$. So the number of different ways of awarding the prizes is $|A \times A \times A \times A|$, which is $|A| \cdot |A| \cdot |A| \cdot |A|$. This is $|A|^4$, or n^4, the same answer we got before.[5]

[5]This discussion assumes that one person can receive any number of prizes. What if the prizes have to go to four different people? This question takes us a little farther into combinatorics than I would like to go, but the answer is not hard. The first award can be given to any of n people. The second prize goes to one of the remaining $n - 1$ people. There are $n - 2$ choices for the third prize and $n - 3$ for the fourth. The number of different ways of awarding the prizes to four different people is the product $n(n-1)(n-2)(n-3)$.

So far, we have only discussed finite sets. $\mathbb{N}$, the set of natural numbers $\{0, 1, 2, 3, \dots\}$, is an example of an infinite set. There is no one-to-one correspondence between $\mathbb{N}$ and any of the finite sets N_n. Another example of an infinite set is the set of even natural numbers, $E = \{0, 2, 4, 6, 8, \dots\}$. There is a natural sense in which the sets $\mathbb{N}$ and E have the same number of elements. That is, there is a one-to-one correspondence between them. The function $f\colon \mathbb{N} \to E$ defined by $f(n) = 2n$ is bijective. We will say that $\mathbb{N}$ and E have the same cardinality, even though that cardinality is not a finite number. Note that E is a proper subset of $\mathbb{N}$. That is, $\mathbb{N}$ has a proper subset that has the same cardinality as $\mathbb{N}$.

We will see that not all infinite sets have the same cardinality. When it comes to infinite sets, intuition is not always a good guide. Most people seem to be torn between two conflicting ideas. On the one hand, they think, it seems that a proper subset of a set should have fewer elements than the set itself. On the other hand, it seems that any two infinite sets should have the same number of elements. Neither of these is true, at least if we define having the same number of elements in terms of one-to-one correspondence.

A set A is said to be ***countably infinite*** if there is a one-to-one correspondence between $\mathbb{N}$ and A. A set is said to be ***countable*** if it is either finite or countably infinite. An infinite set that is not countably infinite is said to be ***uncountable***. If X is an uncountable set, then there is no one-to-one correspondence between $\mathbb{N}$ and X.

The idea of "countable infinity" is that even though a countably infinite set cannot be counted in a finite time, we can imagine counting all the elements of A, one-by-one, in an infinite process. A bijective function $f\colon \mathbb{N} \to A$ provides such an infinite listing: $(f(0), f(1), f(2), f(3), \dots)$. Since f is onto, this infinite list includes all the elements of A. In fact, making such a list effectively shows that A is countably infinite, since the list amounts to a bijective function from $\mathbb{N}$ to A. For an uncountable set, it is impossible to make a list, even an infinite list, that contains all the elements of the set.

Before you start believing in uncountable sets, you should ask for an example. In Chapter 1, we worked with the infinite sets $\mathbb{Z}$ (the integers), $\mathbb{Q}$ (the rationals), $\mathbb{R}$ (the reals), and $\mathbb{R} \smallsetminus \mathbb{Q}$ (the irrationals). Intuitively, these are all "bigger" than $\mathbb{N}$, but as we have already mentioned, intuition is a poor guide when it comes to infinite sets. Are any of $\mathbb{Z}$, $\mathbb{Q}$, $\mathbb{R}$, and $\mathbb{R} \smallsetminus \mathbb{Q}$ in fact uncountable?

It turns out that both $\mathbb{Z}$ and $\mathbb{Q}$ are only countably infinite. The proof that $\mathbb{Z}$ is countable is left as an exercise; we will show here that the set of non-negative rational numbers is countable. (The fact that $\mathbb{Q}$ itself is countable follows easily from this.) The reason is that it's possible to make an infinite list containing all the non-negative rational numbers. Start the list with all the non-negative rational numbers n/m such that $n+m=1$. There is only one such number, namely $0/1$. Next come numbers with $n+m=2$. They are $0/2$ and $1/1$, but we leave out $0/2$ since it's just another way of writing $0/1$, which is already in the list. Now, we add the numbers with $n+m=3$, namely $0/3$, $1/2$, and $2/1$. Again, we leave out $0/3$, since it's equal to a number already in the list. Next come numbers with $n+m=4$. Leaving out $0/4$ and $2/2$ since they are already in the list, we add $1/3$ and $3/1$ to the list. We continue in this way, adding numbers with $n+m=5$, then numbers with $n+m=6$, and so on. The list looks like

$$\left(\frac{0}{1}, \frac{1}{1}, \frac{1}{2}, \frac{2}{1}, \frac{1}{3}, \frac{3}{1}, \frac{1}{4}, \frac{2}{3}, \frac{3}{2}, \frac{4}{1}, \frac{1}{5}, \frac{5}{1}, \frac{1}{6}, \frac{2}{5}, \dots\right)$$

This process can be continued indefinitely, and every non-negative rational number will eventually show up in the list. So we get a complete, infinite list of non-negative rational numbers. This shows that the set of non-negative rational numbers is in fact countable.

On the other hand, $\mathbb{R}$ is uncountable. It is not possible to make an infinite list that contains every real number. It is not even possible to make a list that contains every real number between zero and one. Another way of saying this is that every infinite list of real numbers between zero and one, no matter how it is constructed, leaves something out. To see why this is true, imagine such a list, displayed in an infinitely long column. Each row contains one number, which has an infinite number of digits after the decimal point. Since it is a number between zero and one, the only digit before the decimal point is zero. For example, the list might look like this:

0.**9**0398937249879561297927654857945...
0.1**2**349342094059875980239230834549...
0.22**4**00043298436234709323279989579...
0.500**0**0000000000000000000000000000...
0.7774**3**449234234876990120909480009...
0.77755**5**55588888889498888980000111...
0.123456**7**8888888888888888800000000...
0.3483544**0**009848712712123940320577...
0.93473244**4**47900498340999990948900...
⋮

This is only (a small part of) one possible list. How can we be certain that *every* such list leaves out some real number between zero and one? The trick is to look at the digits shown in bold face. We can use these digits to build a number that is not in the list. Since the first number in the list has a 9 in the first position after the decimal point, we know that this number cannot equal any number of, for example, the form 0.4.... Since the second number has a 2 in the second position after the decimal point, *neither* of the first two numbers in the list is equal to any number that begins with 0.44.... Since the third number has a 4 in the third position after the decimal point, *none* of the first three numbers in the list is equal to any number that begins 0.445.... We can continue to construct a number in this way, and we end up with a number that is different from every number in the list. The nth digit of the number we are building must differ from the nth digit of the nth number in the list. These are the digits shown in bold face in the above list. To be definite, I use a 5 when the corresponding boldface number is 4, and otherwise I use a 4. For the list shown above, this gives a number that begins 0.44544445.... The number constructed in this way is not in the given list, so the list is incomplete. The same construction clearly works for any list of real numbers between zero and one. No such list can be a complete listing of the real numbers between zero and one, and so there can be no complete listing of all real numbers. We conclude that the set $\mathbb{R}$ is uncountable.

The technique used in this argument is called ***diagonalization***. It is named after the fact that the bold face digits in the above list lie along a diagonal line. This proof was discovered by a mathematician named Georg Cantor, who caused quite a fuss in the nineteenth century when he came up with the idea that there are different kinds of infinity. Since then, his notion of using one-to-one correspondence to define the cardinalities of infinite sets has been accepted. Mathematicians now consider it almost intuitive that $\mathbb{N}$, $\mathbb{Z}$, and $\mathbb{Q}$ have the same cardinality while $\mathbb{R}$ has a strictly larger cardinality.

Theorem 2.9. *Suppose that X is an uncountable set, and that K is a countable subset of X. Then the set $X \smallsetminus K$ is uncountable.*

Proof. Let X be an uncountable set. Let $K \subseteq X$, and suppose that K is countable. Let $L = X \smallsetminus K$. We want to show that L is uncountable. Suppose that L is countable. We will show that this assumption leads to a contradiction.

Note that $X = K \cup (X \smallsetminus K) = K \cup L$. You will show in Exercise 11 of this section that the union of two countable sets is countable. Since X is the union of the countable sets K and L, it follows that X is countable. But this contradicts the fact that X is uncountable. This contradiction proves the theorem. □

In the proof, both q and $\neg q$ are shown to follow from the assumptions, where q is the statement "X is countable." The statement q is shown to follow from the assumption that $X \setminus K$ is countable. The statement $\neg q$ is true by assumption. Since q and $\neg q$ cannot both be true, at least one of the assumptions must be false. The only assumption that can be false is the assumption that $X \setminus K$ is countable.

This theorem, by the way, has the following easy corollary. (A ***corollary*** is a theorem that follows easily from another, previously proved theorem.)

Corollary 2.10. *The set of irrational real numbers is uncountable.*

Proof. Let I be the set of irrational real numbers. By definition, $I = \mathbb{R} \setminus \mathbb{Q}$. We have already shown that $\mathbb{R}$ is uncountable and that $\mathbb{Q}$ is countable, so the result follows immediately from the previous theorem. □

You might still think that $\mathbb{R}$ is as big as things get, that is, that any infinite set is in one-to-one correspondence with $\mathbb{R}$ or with some subset of $\mathbb{R}$. In fact, though, if X is any set then it's possible to find a set that has strictly larger cardinality than X. In fact, $\mathcal{P}(X)$ is such a set. A variation of the diagonalization technique can be used to show that there is no one-to-one correspondence between X and $\mathcal{P}(X)$. Note that this is obvious for finite sets, since for a finite set X, $|\mathcal{P}(X)| = 2^{|X|}$, which is larger than $|X|$. The point of the theorem is that it is true even for infinite sets.

Theorem 2.11. *Let X be any set. Then there is no one-to-one correspondence between X and $\mathcal{P}(X)$.*

Proof. Given an arbitrary function $f \colon X \to \mathcal{P}(X)$, we can show that f is not onto. Since a one-to-one correspondence is both one-to-one and onto, this shows that f is not a one-to-one correspondence.

Recall that $\mathcal{P}(X)$ is the set of subsets of X. So, for each $x \in X$, $f(x)$ is a subset of X. We have to show that no matter how f is defined, there is some subset of X that is not in the image of f.

Given f, we define A to be the set $A = \{x \in X \mid x \notin f(x)\}$. The test "$x \notin f(x)$" makes sense because $f(x)$ is a set. Since $A \subseteq X$, we have that $A \in \mathcal{P}(X)$. However, A is not in the image of f. That is, for every $y \in X$, $A \neq f(y)$.[6] To see why this is true, let y be any element of X. There are two cases to consider. Either $y \in f(y)$ or $y \notin f(y)$. We show that whichever case holds, $A \neq f(y)$. If it is true that $y \in f(y)$, then by the definition of A, $y \notin A$. Since $y \in f(y)$ but $y \notin A$, $f(y)$ and A do not have the same elements and therefore are not equal. On the other hand, suppose that $y \notin f(y)$.

[6]In fact, we have constructed A so that the sets A and $f(y)$ differ in at least one element, namely y itself. This is where the "diagonalization" comes in.

Again, by the definition of A, this implies that $y \in A$. Since $y \notin f(y)$ but $y \in A$, $f(y)$ and A do not have the same elements and therefore are not equal. In either case, $A \neq f(y)$. Since this is true for any $y \in X$, we conclude that A is not in the image of f and therefore f is not a one-to-one correspondence. □

From this theorem, it follows that there is no one-to-one correspondence between $\mathbb{R}$ and $\mathcal{P}(\mathbb{R})$. The cardinality of $\mathcal{P}(\mathbb{R})$ is strictly bigger than the cardinality of $\mathbb{R}$. But it doesn't stop there. $\mathcal{P}(\mathcal{P}(\mathbb{R}))$ has an even bigger cardinality, and the cardinality of $\mathcal{P}(\mathcal{P}(\mathcal{P}(\mathbb{R})))$ is bigger still. We could go on like this forever, and we still won't have exhausted all the possible cardinalities. If we let $\mathbb{X}$ be the infinite union $\mathbb{R} \cup \mathcal{P}(\mathbb{R}) \cup \mathcal{P}(\mathcal{P}(\mathbb{R})) \cup \cdots$, then $\mathbb{X}$ has larger cardinality than any of the sets in the union. And then there's $\mathcal{P}(\mathbb{X})$, $\mathcal{P}(\mathcal{P}(\mathbb{X}))$, $\mathbb{X} \cup \mathcal{P}(\mathbb{X}) \cup \mathcal{P}(\mathcal{P}(\mathbb{X})) \cup \cdots$. There is no end to this. There is no upper limit on possible cardinalities, not even an infinite one! We have counted past infinity.

We have seen that $|\mathbb{R}|$ is strictly larger than $|\mathbb{N}|$. We end this section with what might look like a simple question: Is there a subset of $\mathbb{R}$ that is neither in one-to-one correspondence with $\mathbb{N}$ nor with $\mathbb{R}$? That is, is the cardinality of $\mathbb{R}$ the *next* largest cardinality after the cardinality of $\mathbb{N}$, or are there other cardinalities intermediate between them? This problem was unsolved for quite a while, and the solution, when it was found, proved to be completely unexpected. It was shown that both "yes" and "no" are consistent answers to this question! That is, the logical structure built on the system of axioms that had been accepted as the basis of set theory was not extensive enough to answer the question. It is possible to extend the system in various ways. In some extensions, the answer is yes. In others, the answer is no. You might object, "Yes, but which answer is true for the *real* real numbers?" Unfortunately, it's not even clear whether this question makes sense, since in the world of mathematics, the real numbers are just part of a structure built from a system of axioms. And it's not at all clear whether the "real numbers" exist in some sense in the real world. If all this sounds like it's a bit of a philosophical muddle, it is. That's the state of things today at the foundation of mathematics.

Exercises

1. Suppose that A, B, and C are finite sets which are pairwise disjoint. (That is, $A \cap B = A \cap C = B \cap C = \emptyset$.) Express the cardinality of each of the following sets in terms of $|A|$, $|B|$, and $|C|$. Which of your answers depend on the fact that the sets are pairwise disjoint?

a) $\mathcal{P}(A \cup B)$ **b)** $A \times (B^C)$ **c)** $\mathcal{P}(A) \times \mathcal{P}(C)$

d) $A^{B \times C}$ **e)** $(A \times B)^C$ **f)** $\mathcal{P}(A^B)$
g) $(A \cup B)^C$ **h)** $(A \cup B) \times A$ **i)** $A \times A \times B \times B$

2. Suppose that A and B are finite sets which are not necessarily disjoint. What are all the possible values for $|A \cup B|$?

3. Let's say that an "identifier" consists of one or two characters. The fist character is one of the twenty-six letters (A, B, ..., C). The second character, if there is one, is either a letter or one of the ten digits (0, 1, ..., 9). How many different identifiers are there? Explain your answer in terms of unions and cross products.

4. Suppose that there are five books that you might bring along to read on your vacation. In how many different ways can you decide which books to bring, assuming that you want to bring at least one? Why?

5. Show that the cardinality of a finite set is well-defined. That is, show that if f is a bijective function from a set A to N_n, and if g is a bijective function from A to N_m, then $n = m$.

6. Finish the proof of Theorem 2.7 by proving the following statement: Let A be a non-empty set, and let $x \in A$. Let $B = A \smallsetminus \{x\}$. Let $X = \{C \subseteq A \,|\, x \in C\}$. Define $f\colon \mathcal{P}(B) \to X$ by the formula $f(C) = C \cup \{x\}$. Show that f is a bijective function.

7. Use induction on the cardinality of B to show that for any finite sets A and B, $\left|A^B\right| = |A|^{|B|}$. (Hint: For the case where $B \neq \emptyset$, choose $x \in B$, and divide A^B into classes according to the value of $f(x)$.)

8. Let A and B be finite sets with $|A| = n$ and $|B| = m$. List the elements of B as $B = \{b_0, b_1, \ldots, b_{m-1}\}$. Define the function $\mathcal{F}\colon A^B \to A \times A \times \cdots \times A$, where A occurs m times in the cross product, by $\mathcal{F}(f) = \big(f(b_0), f(b_1), \ldots, f(b_{m-1})\big)$. Show that $\mathcal{F}$ is a one-to-one correspondence.

9. Show that $\mathbb{Z}$, the set of integers, is countable by finding a one-to-one correspondence between $\mathbb{N}$ and $\mathbb{Z}$.

10. Show that the set $\mathbb{N} \times \mathbb{N}$ is countable.

11. Complete the proof of Theorem 2.9 as follows:

a) Suppose that A and B are countably infinite sets. Show that $A \cup B$ is countably infinite.

b) Suppose that A and B are countable sets. Show that $A \cup B$ is countable.

12. Prove that each of the following statements is true. In each case, use a proof by contradiction.

a) Let X be a countably infinite set, and let N be a finite subset of X. Then $X \smallsetminus N$ is countably infinite.

b) Let A be an infinite set, and let X be a subset of A. Then at least one of the sets X and $A \smallsetminus X$ is infinite.

c) Every subset of a finite set is finite.

13. Let A and B be sets and let $\bot$ be an entity that is *not* a member of B. Show that there is a one-to-one correspondence between the set of functions from

A to $B \cup \{\perp\}$ and the set of partial functions from A to B. (Partial functions were defined in Section 2.5. The symbol "$\perp$" is sometimes used in theoretical computer science to represent the value "undefined.")

2.7 Relations

In Section 2.4, we saw that "mother of" is a functional relationship because every person has one and only one mother, but that "child of" is not a functional relationship, because a person can have no children or more than one child. However, the relationship expressed by "child of" is certainly one that we have a right to be interested in and one that we should be able to deal with mathematically.

There are many examples of relationships that are not functional relationships. The relationship that holds between two natural numbers n and m when $n \le m$ is an example in mathematics. The relationship between a person and a book that that person has on loan from the library is another. Some relationships involve more than two entities, such as the relationship that associates a name, an address, and a phone number in an address book or the relationship that holds among three real numbers x, y, and z if $x^2 + y^2 + z^2 = 1$. Each of these relationships can be represented mathematically by what is called a "relation."

A ***relation*** on two sets, A and B, is defined to be a subset of $A \times B$. Since a function from A to B is defined, formally, as a subset of $A \times B$ that satisfies certain properties, a function is a relation. However, relations are more general than functions, since *any* subset of $A \times B$ is a relation. We also define a relation among three or more sets to be a subset of the cross product of those sets. In particular, a relation on A, B, and C is a subset of $A \times B \times C$.

For example, if P is the set of people and B is the set of books owned by a library, then we can define a relation $\mathcal{R}$ on the sets P and B to be the set $\mathcal{R} = \{(p, b) \in P \times B \mid p \text{ has } b \text{ out on loan}\}$. The fact that a particular $(p, b) \in \mathcal{R}$ is a fact about the world that the library will certainly want to keep track of. When a collection of facts about the world is stored on a computer, it is called a database. We'll see in the next section that relations are the most common means of representing data in databases.

If A is a set and $\mathcal{R}$ is a relation on the sets A and A (that is, on two copies of A), then $\mathcal{R}$ is said to be a ***binary relation*** on A. That is, a binary relation on the set A is a subset of $A \times A$. The relation consisting of all ordered pairs (c, p) of people such that c is a child of p is a binary relation on the set of people. The set $\{(n, m) \in \mathbb{N} \times \mathbb{N} \mid n \le m\}$ is a binary relation on $\mathbb{N}$. Similarly, we define a ***ternary relation*** on a set A to be a subset of $A \times A \times A$. The set $\{(x, y, z) \in \mathbb{R} \times \mathbb{R} \times \mathbb{R} \mid x^2 + y^2 + z^2 = 1\}$ is a ternary relation on $\mathbb{R}$. For complete generality, we can define an ***n-ary relation*** on A, for any positive integer n, to be a subset of $A \times A \times \cdots \times A$, where A occurs n times in the cross product.

For the rest of this section, we will be working exclusively with binary relations. Suppose that $\mathcal{R} \subseteq A \times A$. That is, suppose that $\mathcal{R}$ is a binary relation on a set A. If $(a, b) \in \mathcal{R}$, then we say that a is related to b by $\mathcal{R}$. Instead of writing "$(a, b) \in \mathcal{R}$", we will often write "$a \,\mathcal{R}\, b$". This notation is used in analogy to the notation $n \le m$ to express the relation that n is less than or equal to m. Remember that $a \,\mathcal{R}\, b$ is just an alternative way of writing $(a, b) \in \mathcal{R}$. In fact, we could consider the relation $\le$ to be a set of ordered pairs and write $(n, m) \in \le$ in place of the notation $n \le m$.

In many applications, attention is restricted to relations that satisfy some property or set of properties. (This is, of course, just what we do when we study functions.) We begin our discussion of binary relations by considering several important properties. In this discussion, let A be a set and let $\mathcal{R}$ be a binary relation on A, that is, a subset of $A \times A$.

$\mathcal{R}$ is said to be ***reflexive*** if $\forall a \in A\,(a \,\mathcal{R}\, a)$. That is, a binary relation on a set is reflexive if every element of the set is related to itself. This is true, for example, for the relation $\le$ on the set $\mathbb{N}$, since $n \le n$ for every $n \in \mathbb{N}$. On the other hand, it is not true for the relation $<$ on $\mathbb{N}$, since, for example, the statement $17 < 17$ is false.[7]

$\mathcal{R}$ is called ***transitive*** if $\forall a \in A$, $\forall b \in A$, $\forall c \in A\,\big((a \,\mathcal{R}\, b \wedge b \,\mathcal{R}\, c) \rightarrow (a \,\mathcal{R}\, c)\big)$. Transitivity allows us to "chain together" two true statements $a \,\mathcal{R}\, b$ and $b \,\mathcal{R}\, c$, which are "linked" by the b that occurs in each statement, to deduce that $a \,\mathcal{R}\, c$. For example, suppose P is the set of people, and define the relation $\mathcal{C}$ on P such that $x \,\mathcal{P}\, y$ if and only if x is a child of y. The relation $\mathcal{P}$ is not transitive because the child of a child of a person is not a child of that person. Suppose, on the other hand, that we define a relation $\mathcal{D}$ on P such that $x \,\mathcal{D}\, y$ if and only if x is a descendent of y. Then D is a transitive relation on the set of people, since a descendent of a descendent of a person is a descendent of that person. That is, from the facts that Elizabeth is a descendent of Victoria and Victoria is a descendent of James, we can deduce that Elizabeth is a descendent of James. In the mathematical world, the relations $\le$ and $<$ on the set $\mathbb{N}$ are both transitive.

$\mathcal{R}$ is said to be ***symmetric*** if $\forall a \in A$, $\forall b \in B\,(a \,\mathcal{R}\, b \rightarrow b \,\mathcal{R}\, a)$. That is, whenever a is related to b, it follows that b is related to a. The relation "is

[7]Note that to show that the relation $\mathcal{R}$ is *not* reflexive, you only need to find one a such that $a \,\mathcal{R}\, a$ is false. This follows from the fact that $\neg\big(\forall a \in A\,(a \,\mathcal{R}\, a)\big) \equiv \exists a \in A\,\big(\neg(a \,\mathcal{R}\, a)\big)$. A similar remark holds for each of the properties of relations that are discussed here.

a first cousin of" on the set of people is symmetric, since whenever x is a first cousin of y, we have automatically that y is a first cousin of x. On the other hand, the "child of" relation is certainly not symmetric. The relation $\leq$ on $\mathbb{N}$ is not symmetric. From the fact that $n \leq m$, we cannot conclude that $m \leq n$. It is true for *some* n and m in $\mathbb{N}$ that $n \leq m \rightarrow m \leq n$, but it is not true for *all* n and m in $\mathbb{N}$.

Finally, $\mathcal{R}$ is ***antisymmetric*** if $\forall a \in A, \forall b \in B\ \big((a\,\mathcal{R}\,b \wedge b\,\mathcal{R}\,a) \rightarrow a = b\big)$. The relation $\mathcal{R}$ is antisymmetric if for any two *distinct* elements x and y of A, we can't have both $x\,\mathcal{R}\,y$ and $y\,\mathcal{R}\,x$. The relation $\leq$ on $\mathbb{N}$ is antisymmetric because from the facts that $n \leq m$ and $m \leq n$, we can deduce that $n = m$. The relation "child of" on the set of people is antisymmetric since it's impossible to have both that x is a child of y and y is a child of x.

There are a few combinations of properties that define particularly useful types of binary relations. The relation $\leq$ on the set $\mathbb{N}$ is reflexive, antisymmetric, and transitive. These properties define what is called a partial order: A ***partial order*** on a set A is a binary relation on A that is reflexive, antisymmetric, and transitive.

Another example of a partial order is the subset relation, $\subseteq$, on the power set of any set. If X is a set, then of course $\mathcal{P}(X)$ is a set in its own right, and $\subseteq$ can be considered to be a binary relation on this set. Two elements A and B of $\mathcal{P}(X)$ are related by $\subseteq$ if and only if $A \subseteq B$. This relation is reflexive since every set is a subset of itself. The fact that it is antisymmetric follows from Theorem 2.1. The fact that it is transitive was Exercise 11 in Section 2.1.

The ordering imposed on $\mathbb{N}$ by $\leq$ has one important property that the ordering of subsets by $\subseteq$ does not share. If n and m are natural numbers, then at least one of the statements $n \leq m$ and $m \leq n$ must be true. However, if A and B are subsets of a set X, it is certainly possible that both $A \subseteq B$ and $B \subseteq A$ are false. A binary relation $\mathcal{R}$ on a set A is said to be a ***total order*** if it is a partial order and furthermore for any two elements a and b of A, either $a\,\mathcal{R}\,b$ or $b\,\mathcal{R}\,a$. The relation $\leq$ on the set $\mathbb{N}$ is a total order. The relation $\subseteq$ on $\mathcal{P}(X)$ is not. (Note once again the slightly odd mathematical language: A total order is a kind of partial order—not, as you might expect, the opposite of a partial order.)

For another example of ordering, let L be the set of strings that can be made from lowercase letters. L contains both English words and nonsense strings such as "sxjja". There is a commonly used total order on the set L, namely alphabetical order.

We'll approach another important kind of binary relation indirectly, through what might at first appear to be an unrelated idea. Let A be a set. A ***partition*** of A is defined to be a collection of non-empty subsets of A such that each pair of distinct subsets in the collection is disjoint and the union of all the subsets in the collection is A. A partition of A is just a division of all the elements of A into non-overlapping subsets. For example, the sets $\{1, 2, 6\}$, $\{3, 7\}$, $\{4, 5, 8, 10\}$, and $\{9\}$ form a partition of the set $\{1, 2, \ldots, 10\}$. Each element of $\{1, 2, \ldots, 10\}$ occurs in exactly one of the sets that make up the partition. As another example, we can partition the set of all people into two sets, the set of males and the set of females. Biologists try to partition the set of all organisms into different species. Librarians try to partition books into various categories such as fiction, biography, and poetry. In the real world, classifying things into categories is an essential activity, although the boundaries between categories are not always well-defined. The abstract mathematical notion of a partition of a set models the real-world notion of classification. In the mathematical world, though, the categories are sets and the boundary between two categories is sharp.

In the real world, items are classified in the same category because they are related in some way. This leads us from partitions back to relations. Suppose that we have a partition of a set A. We can define a relation $\mathcal{R}$ on A by declaring that for any a and b in A, $a \mathcal{R} b$ if and only if a and b are members of the same subset in the partition. That is, two elements of A are related if they are in the same category. It is clear that the relation defined in this way is reflexive, symmetric, and transitive.

An ***equivalence relation*** is defined to be a binary relation that is reflexive, symmetric, and transitive. Any relation defined, as above, from a partition is an equivalence relation. Conversely, we can show that any equivalence relation defines a partition. Suppose that $\mathcal{R}$ is an equivalence relation on a set A. Let $a \in A$. We define the ***equivalence class*** of a under the equivalence relation $\mathcal{R}$ to be the subset $[a]_{\mathcal{R}}$ defined as $[a]_{\mathcal{R}} = \{b \in A \mid b \mathcal{R} a\}$. That is, the equivalence class of a is the set of all elements of A that are related to a. In most cases, we'll assume that the relation in question is understood, and we'll write $[a]$ instead of $[a]_{\mathcal{R}}$. Note that each equivalence class is a subset of A. The following theorem shows that the collection of equivalence classes form a partition of A.

Theorem 2.12. *Let A be a set and let $\mathcal{R}$ be an equivalence relation on A. Then the collection of all equivalence classes under $\mathcal{R}$ is a partition of A.*

Proof. To show that a collection of subsets of A is a partition, we must show that each subset is non-empty, that the intersection of two distinct subsets is empty, and that the union of all the subsets is A.

If $[a]$ is one of the equivalence classes, it is certainly non-empty, since

$a \in [a]$. (This follows from the fact that $\mathcal{R}$ is reflexive, and hence $a \,\mathcal{R}\, a$.) To show that A is the union of all the equivalence classes, we just have to show that each element of A is a member of one of the equivalence classes. Again, the fact that $a \in [a]$ for each $a \in A$ shows that this is true.

Finally, we have to show that the intersection of two distinct equivalence classes is empty. Suppose that a and b are elements of A and consider the equivalence classes $[a]$ and $[b]$. We have to show that if $[a] \neq [b]$, then $[a] \cap [b] = \emptyset$. Equivalently, we can show the converse: If $[a] \cap [b] \neq \emptyset$ then $[a] = [b]$. So, assume that $[a] \cap [b] \neq \emptyset$. Saying that a set is not empty just means that the set contains some element, so there must be an $x \in A$ such that $x \in [a] \cap [b]$. Since $x \in [a]$, $x \,\mathcal{R}\, a$. Since $\mathcal{R}$ is symmetric, we also have $a \,\mathcal{R}\, x$. Since $x \in [b]$, $x \,\mathcal{R}\, b$. Since $\mathcal{R}$ is transitive and since $(a \,\mathcal{R}\, x) \wedge (x \,\mathcal{R}\, b)$, it follows that $a \,\mathcal{R}\, b$.

Our object is to deduce that $[a] = [b]$. Since $[a]$ and $[b]$ are sets, they are equal if and only if $[a] \subseteq [b]$ and $[b] \subseteq [a]$. To show that $[a] \subseteq [b]$, let c be an arbitrary element of $[a]$. We must show that $c \in [b]$. Since $c \in [a]$, we have that $c \,\mathcal{R}\, a$. And we have already shown that $a \,\mathcal{R}\, b$. From these two facts and the transitivity of $\mathcal{R}$, it follows that $c \,\mathcal{R}\, b$. By definition, this means that $c \in [b]$. We have shown that any member of $[a]$ is a member of $[b]$ and therefore that $[a] \subseteq [b]$. The fact that $[b] \subseteq [a]$ can be shown in the same way. We deduce that $[a] = [b]$, which proves the theorem. □

The point of this theorem is that if we can find a binary relation that satisfies certain properties, namely the properties of an equivalence relation, then we can classify things into categories, where the categories are the equivalence classes.

For example, suppose that U is a possibly infinite set. Define a binary relation $\sim$ on $\mathcal{P}(U)$ as follows: For X and Y in $\mathcal{P}(U)$, $X \sim Y$ if and only if there is a bijective function from the set X to the set Y. In other words, $X \sim Y$ means that X and Y have the same cardinality. Then $\sim$ is an equivalence relation on $\mathcal{P}(U)$. (The symbol $\sim$ is often used to denote equivalence relations. It is usually read "is equivalent to.") If $X \in \mathcal{P}(U)$, then the equivalence class $[X]_\sim$ consists of all the subsets of U that have the same cardinality as X. We have classified all the subsets of U according to their cardinality—even though we have never said what an infinite cardinality *is*. (We have only said what it means to have the same cardinality.)

You might remember a popular puzzle called Rubic's Cube, a cube made of smaller cubes with colored sides that could be manipulated by twisting layers of little cubes. The object was to manipulate the cube so that the colors of the little cubes formed a certain configuration. Define two configurations of the cube to be equivalent if it's possible to manipulate one configuration into the other by a sequence of twists. This is, in fact, an equivalence relation on the set of possible configurations. (Symmetry follows from the fact that each move is reversible.) It has been shown that this equivalence relation has exactly twelve equivalence classes. The interesting fact is that it has more than one equivalence class: If the configuration that the cube is in and the configuration that you want to achieve are not in the same equivalence class, then you are doomed to failure.

Suppose that $\mathcal{R}$ is a binary relation on a set A. Even though $\mathcal{R}$ might not be transitive, it is always possible to construct a transitive relation from $\mathcal{R}$ in a natural way. If we think of $a \,\mathcal{R}\, b$ as meaning that a is related by $\mathcal{R}$ to b "in one step," then we consider the relationship that holds between two elements x and y when x is related by $\mathcal{R}$ to y "in one or more steps." This relationship defines a binary relation on A that is called the ***transitive closure*** of $\mathcal{R}$. The transitive closure of $\mathcal{R}$ is denoted $\mathcal{R}^*$. Formally, $\mathcal{R}^*$ is defined as follows: For a and b in A, $a \,\mathcal{R}^*\, b$ if there is a sequence $x_0, x_1, \dots x_n$ of elements of A, where $n > 0$ and $x_0 = a$ and $x_n = b$, such that $x_0 \,\mathcal{R}\, x_1$, $x_1 \,\mathcal{R}\, x_2$, ..., and $x_{n-1} \,\mathcal{R}\, x_n$.

For example, if $a \,\mathcal{R}\, c$, $c \,\mathcal{R}\, d$, and $d \,\mathcal{R}\, b$, then we would have that $a \,\mathcal{R}^*\, b$. Of course, we would also have that $a \,\mathcal{R}^*\, c$, and $a \,\mathcal{R}^*\, d$.

For a practical example, suppose that C is the set of all cities and let $\mathcal{A}$ be the binary relation on C such that for x and y in C, $x \,\mathcal{A}\, y$ if there is a regularly scheduled airline flight from x to y. Then the transitive closure $\mathcal{A}^*$ has a natural interpretation: $x \,\mathcal{A}^*\, y$ if it's possible to get from x to y by a sequence of one or more regularly scheduled airline flights. You'll find a few more examples of transitive closures in the exercises.

Exercises

1. For a finite set, it is possible to define a binary relation on the set by listing the elements of the relation, considered as a set of ordered pairs. Let A be the set $\{a, b, c, d\}$, where a, b, c, and d are distinct. Consider each of the following binary relations on A. Is the relation reflexive? Symmetric? Antisymmetric? Transitive? Is it a partial order? An equivalence relation?

a) $\mathcal{R} = \{(a, b), (a, c), (a, d)\}$.

b) $\mathcal{S} = \{(a, a), (b, b), (c, c), (d, d), (a, b), (b, a)\}$.

c) $\mathcal{T} = \{(b, b), (c, c), (d, d)\}$.

d) $\mathcal{C} = \{(a, b), (b, c), (a, c), (d, d)\}$.

e) $\mathcal{D} = \{(a, b), (b, a), (c, d), (d, c)\}$.

2. Let A be the set $\{1, 2, 3, 4, 5, 6\}$. Consider the partition of A into the subsets $\{1, 4, 5\}$, $\{3\}$, and $\{2, 6\}$. Write out the associated equivalence relation on A as a set of ordered pairs.

3. Consider each of the following relations on the set of people. Is the relation reflexive? Symmetric? Transitive? Is it an equivalence relation?

a) x is related to y if x and y have the same biological parents.
b) x is related to y if x and y have at least one biological parent in common.
c) x is related to y if x and y were born in the same year.
d) x is related to y if x is taller than y.
e) x is related to y if x and y have both visited Honolulu.

4. It is possible for a relation to be both symmetric and antisymmetric. For example, the equality relation, $=$, is a relation on any set which is both symmetric and antisymmetric. Suppose that A is a set and $\mathcal{R}$ is a relation on A that is both symmetric and antisymmetric. Show that $\mathcal{R}$ is a subset of $=$ (when both relations are considered as sets of ordered pairs). That is, show that for any a and b in A, $(a\,\mathcal{R}\,b) \rightarrow (a = b)$.

5. Let $\sim$ be the relation on $\mathbb{R}$, the set of real numbers, such that for x and y in $\mathbb{R}$, $x \sim y$ if and only if $x - y \in \mathbb{Z}$. For example, $\sqrt{2} - 1 \sim \sqrt{2} + 17$ because the difference, $(\sqrt{2} - 1) - (\sqrt{2} + 17)$, is -18, which is an integer. Show that $\sim$ is an equivalence relation. Show that each equivalence class $[x]_\sim$ contains exactly one number a which satisfies $0 \le a < 1$. (Thus, the set of equivalence classes under $\sim$ is in one-to-one correspondence with the half-open interval $[0, 1)$.)

6. Let A and B be any sets, and suppose $f\colon A \rightarrow B$. Define a relation $\sim$ on B such that for any x and y in A, $x \sim y$ if and only if $f(x) = f(y)$. Show that $\sim$ is an equivalence relation on A.

7. Let $\mathbb{Z}^+$ be the set of positive integers $\{1, 2, 3, \dots\}$. Define a binary relation $\mathcal{D}$ on $\mathbb{Z}^+$ such that for n and m in $\mathbb{Z}^+$, $n\,\mathcal{D}\,m$ if n divides evenly into m, with no remainder. Equivalently, $n\,\mathcal{D}\,m$ if n is a factor of m, that is, if there is a k in $\mathbb{Z}^+$ such that $m = nk$. Show that $\mathcal{D}$ is a partial order.

8. Consider the set $\mathbb{N} \times \mathbb{N}$, which consists of all ordered pairs of natural numbers. Since $\mathbb{N} \times \mathbb{N}$ is a set, it is possible to have binary relations on $\mathbb{N} \times \mathbb{N}$. Such a relation would be a subset of $(\mathbb{N} \times \mathbb{N}) \times (\mathbb{N} \times \mathbb{N})$. Define a binary relation $\preceq$ on $\mathbb{N} \times \mathbb{N}$ such that for (m, n) and (k, ℓ) in $\mathbb{N} \times \mathbb{N}$, $(m, n) \preceq (k, \ell)$ if and only if either $m < k$ or $((m = k) \wedge (n \le \ell))$. Which of the following are true?

a) $(2, 7) \preceq (5, 1)$ **b)** $(8, 5) \preceq (8, 0)$
c) $(0, 1) \preceq (0, 2)$ **d)** $(17, 17) \preceq (17, 17)$

Show that $\preceq$ is a total order on $\mathbb{N} \times \mathbb{N}$.

9. Let $\sim$ be the relation defined on $\mathbb{N} \times \mathbb{N}$ such that $(n, m) \sim (k, \ell)$ if and only if $n + \ell = m + k$. Show that $\sim$ is an equivalence relation.

10. Let P be the set of people and let $\mathcal{C}$ be the "child of" relation. That is $x\,\mathcal{C}\,y$ means that x is a child of y. What is the meaning of the transitive closure $\mathcal{C}^*$? Explain your answer.

11. Let $\mathcal{R}$ be the binary relation on $\mathbb{N}$ such that $x\,\mathcal{R}\,y$ if and only if $y = x + 1$. Identify the transitive closure $\mathcal{R}^*$. (It is a well-known relation.) Explain your answer.

12. Suppose that $\mathcal{R}$ is a reflexive, symmetric binary relation on a set A. Show that the transitive closure $\mathcal{R}^*$ is an equivalence relation.

2.8 Application: Relational Databases

One of the major uses of computer systems is to store and manipulate collections of data. A ***database*** is a collection of data that has been organized so that it is possible to add and delete information, to update the data that it contains, and to retrieve specified parts of the data. A ***Database Management System***, or DBMS, is a computer program that makes it possible to create and manipulate databases. A DBMS must be able to accept and process commands that manipulate the data in the databases that it manages. These commands are called ***queries***, and the languages in which they are written are called ***query languages***. A query language is a kind of specialized programming language.

There are many different ways that the data in a database could be represented. Different DBMS's use various data representations and various query languages. However, data is most commonly stored in relations. A relation in a database is a relation in the mathematical sense. That is, it is a subset of a cross product of sets. A database that stores its data in relations is called a ***relational database***. The query language for most relational database management systems is some form of the language known as ***Structured Query Language***, or SQL. In this section, we'll take a very brief look at SQL, relational databases, and how they use relations.

A relation is just a subset of a cross product of sets. Since we are discussing computer representation of data, the sets are data types. As in Section 2.5, we'll use data type names such as *int* and *string* to refer to these sets. A relation that is a subset of the cross product $int \times int \times string$ would consist of ordered 3-tuples such as (17, 42, "hike"). In a relational database, the data is stored in the form of one or more such relations. The relations are called tables, and the tuples that they contain are called rows or records.

As an example, consider a lending library that wants to store data about its members, the books that it owns, and which books the members have out on loan. This data could be represented in three tables, as illustrated in Figure 2.4. The relations are shown as tables rather than as sets of ordered tuples, but each table is, in fact, a relation. The rows of the table are the tuples. The `Members` table, for example, is a subset of $int \times string \times string \times string$, and one of the tuples is (1782, "Smith, John", "107 Main St", "New York, NY"). A table does have one thing that ordinary relations in mathematics do not have. Each column in the table has a name. These names are used in the query language to manipulate the data in the tables.

The data in the `Members` table is the basic information that the library needs in order to keep track of its members, namely the name and address of each member. A member also has a `MemberID` number, which is presumably assigned by the library. Two different members can't have the same `MemberID`, even though they might have the same name or the same address. The `MemberID` acts as a ***primary key*** for the `Members` table. A given value of the primary key uniquely identifies one of the rows of the table. Similarly, the `BookID` in the `Books` table is a primary key for that table. In the `Loans` table, which holds information about which books are out on loan to which members, a `MemberID` unambiguously identifies the member who has a given book on loan, and the `BookID` says unambiguously which book that is. Every table has a primary key, but the key can consist of more than one column. The DBMS enforces the uniqueness of primary keys. That is, it won't let users make a modification to the table if it would result in two rows having the same primary key.

The fact that a relation is a set—a set of tuples—means that it can't contain the same tuple more than once. In terms of tables, this means that a table shouldn't contain two identical rows. But since no two rows can contain the same primary key, it's impossible for two rows to be identical. So tables are in fact relations in the mathematical sense.

The library must have a way to add and delete members and books and to make a record when a book is borrowed or returned. It should also have a way to change the address of a member or the due date of a borrowed book. Operations such as these are performed using the DBMS's query language. SQL has commands named `INSERT`, `DELETE`, and `UPDATE` for performing these operations. The command for adding Barack Obama as a member of the library with `MemberID` 999 would be

```
INSERT INTO Members
VALUES (999, "Barack Obama",
            "1600 Pennsylvania Ave", "Washington, DC")
```

When it comes to deleting and modifying rows, things become more interesting because it's necessary to specify which row or rows will be affected. This is done by specifying a condition that the rows must fulfill. For example, this command will delete the member with ID 4277:

```
DELETE FROM Members
WHERE MemberID = 4277
```

It's possible for a command to affect multiple rows. For example,

```
DELETE FROM Members
WHERE Name = "Smith, John"
```

would delete every row in which the name is "Smith, John." The update command also specifies what changes are to be made to the row:

```
UPDATE Members
SET Address="19 South St", City="Hartford, CT"
WHERE MemberID = 4277
```

Of course, the library also needs a way of retrieving information from the database. SQL provides the SELECT command for this purpose. For example, the query

```
SELECT Name, Address
FROM Members
WHERE City = "New York, NY"
```

asks for the name and address of every member who lives in New York City. The last line of the query is a condition that picks out certain rows of the "Members" relation, namely all the rows in which the City is "New York, NY". The first line specifies which data from those rows should be retrieved. The data is actually returned in the form of a table. For example, given the data in Figure 2.4, the query would return this table:

Smith, John	107 Main St
Jones, Mary	1515 Center Ave
Lee, Joseph	90 Park Ave
O'Neil, Sally	89 Main St

The table returned by a SELECT query can even be used to construct more complex queries. For example, if the table returned by SELECT has only one column, then it can be used with the IN operator to specify any value listed in that column. The following query will find the BookID of every book that is out on loan to a member who lives in New York City:

```
SELECT BookID
FROM Loans
WHERE MemberID IN (SELECT MemberID
                   FROM Members
                   WHERE City = "New York, NY")
```

More than one table can be listed in the FROM part of a query. The tables that are listed are joined into one large table, which is then used for the query. The large table is essentially the cross product of the joined tables, when the tables are understood as sets of tuples. For example, suppose that we want the titles of all the books that are out on loan to members who live in New York City. The titles are in the `Books` table, while information about loans is in the `Loans` table. To get the desired data, we can join the tables and extract the answer from the joined table:

```
SELECT Title
FROM Books, Loans
WHERE MemberID IN (SELECT MemberID
                   FROM Members
                   WHERE City = "New York, NY")
```

In fact, we can do the same query without using the nested `SELECT`. We need one more bit of notation: If two tables have columns that have the same name, the columns can be named unambiguously by combining the table name with the column name. For example, if the `Members` table and `Loans` table are both under discussion, then the `MemberID` columns in the two tables can be referred to as `Members.MemberID` and `Loans.MemberID`. So, we can say:

```
SELECT Title
FROM Books, Loans
WHERE City ="New York, NY"
      AND Members.MemberID = Loans.MemberID
```

This is just a sample of what can be done with SQL and relational databases. The conditions in `WHERE` clauses can get very complicated, and there are other operations besides the cross product for combining tables. The database operations that are needed to complete a given query can be complex and time-consuming. Before carrying out a query, the DBMS tries to optimize it. That is, it manipulates the query into a form that can be carried out most efficiently. The rules for manipulating and simplifying queries form an *algebra* of relations, and the theoretical study of relational databases is in large part the study of the algebra of relations.

Exercises

1. Using the library database from Figure 2.4, what is the result of each of the following SQL commands?

a)
```
SELECT Name, Address
FROM Members
```

```
WHERE Name = "Smith, John"
```

b)
```
DELETE FROM Books
WHERE Author = "Isaac Asimov"
```

c)
```
UPDATE Loans
SET DueDate = "November 20"
WHERE BookID = 221
```

d)
```
SELECT Title
FROM Books, Loans
WHERE Books.BookID = Loans.BookID
```

e)
```
DELETE FROM Loans
WHERE MemberID IN (SELECT MemberID
                   FROM Members
                   WHERE Name = "Lee, Joseph")
```

2. Using the library database from Figure 2.4, write an SQL command to do each of the following database manipulations:

a) Find the `BookID` of every book that is due on November 1, 2010.

b) Change the `DueDate` of the book with `BookID` 221 to November 15, 2010.

c) Change the `DueDate` of the book with title "Summer Lightning" to November 14, 2010. Use a nested `SELECT`.

d) Find the name of every member who has a book out on loan. Use joined tables in the `FROM` clause of a `SELECT` command.

3. Suppose that a college wants to use a database to store information about its students, the courses that are offered in a given term, and which students are taking which courses. Design tables that could be used in a relational database for representing this data. Then write SQL commands to do each of the following database manipulations. (You should design your tables so that they can support all these commands.)

a) Enroll the student with ID number 1928882900 in "English 260".

b) Remove "John Smith" from "Biology 110".

c) Remove the student with ID number 2099299001 from every course in which that student is enrolled.

d) Find the names and addresses of the students who are taking "Computer Science 229".

e) Cancel the course "History 101".

Members

MemberID	Name	Address	City
1782	Smith, John	107 Main St	New York, NY
2889	Jones, Mary	1515 Center Ave	New York, NY
378	Lee, Joseph	90 Park Ave	New York, NY
4277	Smith, John	2390 River St	Newark, NJ
5704	O'Neil, Sally	89 Main St	New York, NY

Books

BookID	Title	Author
182	I, Robot	Isaac Asimov
221	The Sound and the Fury	William Faulkner
38	Summer Lightning	P.G. Wodehouse
437	Pride and Prejudice	Jane Austen
598	Left Hand of Darkness	Ursula LeGuin
629	Foundation Trilogy	Isaac Asimov
720	Mirror Dance	Lois McMaster Bujold

Loans

MemberID	BookID	DueDate
378	221	October 8, 2010
2889	182	November 1, 2010
4277	221	November 1, 2010
1782	38	October 30, 2010

Figure 2.4: Tables that could be part of a relational database. Each table has a name, shown above the table. Each column in the table also has a name, shown in the top row of the table. The remaining rows hold the data.

Chapter 3

Regular Expressions and Finite-State Automata

WITH THE SET OF MATHEMATICAL TOOLS from the first two chapters, we are now ready to study ***languages*** and ***formal language theory***. Our intent is to examine the question of how, and which, languages can be mechanically generated and recognized; and, ultimately, to see what this tells us about what computers can and can't do.

3.1 Languages

In formal language theory, an ***alphabet*** is a finite, non-empty set. The elements of the set are called ***symbols***. A finite sequence of symbols $a_1a_2\ldots a_n$ from an alphabet is called a ***string*** over that alphabet.

Example 3.1. $\Sigma = \{0,1\}$ is an alphabet, and *011*, *1010*, and *1* are all strings over Σ.

Note that strings really are *sequences* of symbols, which implies that order matters. Thus *011*, *101*, and *110* are all different strings, though they are made up of the same symbols. The strings $x = a_1a_2\ldots a_n$ and $y = b_1b_2\ldots b_m$ are ***equal*** only if $m = n$ (i.e. the strings contain the same number of symbols) and $a_i = b_i$ for all $1 \leq i \leq n$.

Just as there are operations defined on numbers, truth values, sets, and other mathematical entities, there are operations defined on strings. Some important operations are:

1. *length*: the **length** of a string x is the number of symbols in it. The notation for the length of x is $|x|$. Note that this is consistent with other uses of $|\ |$, all of which involve some notion of size: $|number|$ measures how big a number is (in terms of its distance from 0); $|set|$ measures the size of a set (in terms of the number of elements).

   We will occasionally refer to a *length-n string*. This is a slightly awkward, but concise, shorthand for "a string whose length is n".

2. *concatenation*: the **concatenation** of two strings $x = a_1a_2 \ldots a_m$ and $y = b_1b_2 \ldots b_n$ is the sequence of symbols $a_1 \ldots a_mb_1 \ldots b_n$. Sometimes $\cdot$ is used to denote concatenation, but it is far more usual to see the concatenation of x and y denoted by xy than by $x \cdot y$. You can easily convince yourself that concatenation is associative (i.e. $(xy)z = x(yz)$ for all strings x, y and z.) Concatenation is not commutative (i.e. it is not always true that $xy = yx$: for example, if $x = a$ and $y = b$ then $xy = ab$ while $yx = ba$ and, as discussed above, these strings are not equal.)

3. *reversal*: the **reverse** of a string $x = a_1a_2 \ldots a_n$ is the string $x^R = a_na_{n-1} \ldots a_2a_1$.

Example 3.2. Let $\Sigma = \{a, b\}$, $x = a$, $y = abaa$, and $z = bab$. Then $|x| = 1$, $|y| = 4$, and $|z| = 3$. Also, $xx = aa$, $xy = aabaa$, $xz = abab$, and $zx = baba$. Finally, $x^R = a$, $y^R = aaba$, and $z^R = bab$.

By the way, the previous example illustrates a naming convention standard throughout language theory texts: if a letter is intended to represent a single symbol in an alphabet, the convention is to use a letter from the beginning of the English alphabet (a, b, c, d); if a letter is intended to represent a string, the convention is to use a letter from the end of the English alphabet (u, v, etc).

In set theory, we have a special symbol to designate the set that contains no elements. Similarly, language theory has a special symbol ε which is used to represent the ***empty string***, the string with no symbols in it. (Some texts use the symbol λ instead.) It is worth noting that $|\varepsilon| = 0$, that $\varepsilon^R = \varepsilon$, and that $\varepsilon \cdot x = x \cdot \varepsilon = x$ for all strings x. (This last fact may appear a bit confusing. Remember that ε is not a symbol in a string with length 1, but rather the name given to the string made up of 0 symbols. Pasting those 0 symbols onto the front or back of a string x still produces x.)

The set of all strings over an alphabet Σ is denoted Σ^*. (In language theory, the symbol * is typically used to denote "zero or more", so Σ^* is the set of strings made up of zero or more symbols from Σ.) Note that while an alphabet Σ is by definition a *finite* set of symbols, and strings are by definition *finite* sequences of those symbols, the set Σ^* is *always infinite.* Why is this? Suppose Σ contains n elements. Then there is one string over Σ with 0 symbols, n strings with 1 symbol, n^2 strings with 2 symbols (since there are n choices for the first symbol and n choices for the second), n^3 strings with 3 symbols, etc.

Example 3.3. If $\Sigma = \{1\}$, then $\Sigma^* = \{\varepsilon, 1, 11, 111, \ldots\}$. If $\Sigma = \{a, b\}$, then $\Sigma^* = \{\varepsilon, a, b, aa, ab, ba, bb, aaa, aab, \ldots\}$.

Note that Σ^* is *countably* infinite: if we list the strings as in the preceding example (length-0 strings, length-1 strings in "alphabetical" order, length-2 strings similarly ordered, etc) then any string over Σ will eventually appear. (In fact, if $|\Sigma| = n \geq 2$ and $x \in \Sigma^*$ has length k, then x will appear on the list within the first $\frac{n^{k+1}-1}{n-1}$ entries.)

We now come to the definition of a ***language*** in the formal language theoretical sense.

Definition 3.1. A language over an alphabet Σ is a subset of Σ^*. Thus, a language over Σ is an element of $\mathcal{P}(\Sigma^*)$, the power set of Σ^*.

In other words, any set of strings (over alphabet Σ) constitutes a language (over alphabet Σ).

Example 3.4. Let $\Sigma = \{0, 1\}$. Then the following are all languages over Σ:

$L_1 = \{011, 1010, 111\}$

$L_2 = \{0, 10, 110, 1110, 11110, \ldots\}$

$L_3 = \{x \in \Sigma^* \mid n_0(x) = n_1(x)\}$, where the notation $n_0(x)$ stands for the

number of 0's in the string x, and similarly for $n_1(x)$.

$L_4 = \{x \mid x \text{ represents a multiple of 5 in binary}\}$

Note that languages can be either finite or infinite. Because Σ^* is infinite, it clearly has an infinite number of subsets, and so there are an infinite number of languages over Σ. But are there countably or uncountably many such languages?

Theorem 3.1. *For any alphabet Σ, the number of languages over Σ is uncountable.*

This fact is an immediate consequence of the result, proved in a previous chapter, that the power set of a countably infinite set is uncountable. Since the elements of $\mathcal{P}(\Sigma)$ are exactly the languages over Σ, there are uncountably many such languages.

Languages are sets and therefore, as for any sets, it makes sense to talk about the union, intersection, and complement of languages. (When taking the complement of a language over an alphabet Σ, we always consider the univeral set to be Σ^*, the set of all strings over Σ.) Because languages are sets of strings, there are additional operations that can be defined on languages, operations that would be meaningless on more general sets. For example, the idea of concatenation can be extended from strings to languages.

For two sets of strings S and T, we define the ***concatenation*** of S and T (denoted $S \cdot T$ or just ST) to be the set $ST = \{st \mid s \in S \wedge t \in T\}$. For example, if $S = \{ab, aab\}$ and $T = \{\varepsilon, 110, 1010\}$, then $ST = \{ab, ab110, ab1010, aab, aab110, aab1010\}$. Note in particular that $ab \in ST$, because $ab \in S$, $\varepsilon \in T$, and $ab \cdot \varepsilon = ab$. Because concatenation of sets is defined in terms of the concatenation of the strings that the sets contain, concatenation of sets is associative and not commutative. (This can easily be verified.)

When a set S is concatenated with itself, the notation SS is usually scrapped in favour of S^2; if S^2 is concatenated with S, we write S^3 for the resulting set, etc. So S^2 is the set of all strings formed by concatenating two (possibly different, possibly identical) strings from S, S^3 is the set of strings formed by concatenating three strings from S, etc. Extending this notation, we take S^1 to be the set of strings formed from one string in S (i.e. S^1 is S itself), and S^0 to be the set of strings formed from zero strings in S (i.e. $S^0 = \{\varepsilon\}$). If we take the union $S^0 \cup S^1 \cup S^2 \cup \ldots$, then the resulting set is the set of all strings formed by concatenating zero or more strings from S, and is denoted S^*. The set S^* is called the ***Kleene closure*** of S, and the * operator is called the ***Kleene star*** operator.

Example 3.5. Let $S = \{01, ba\}$. Then
$S^0 = \{\varepsilon\}$
$S^1 = \{01, ba\}$
$S^2 = \{0101, 01ba, ba01, baba\}$
$S^3 = \{010101, 0101ba, 01ba01, 01baba, ba0101, ba01ba, baba01, bababa\}$
etc, so
$S^* = \{\varepsilon, 01, ba, 0101, 01ba, ba01, baba, 010101, 0101ba, \ldots\}$.

Note that this is the second time we have seen the notation *something**. We have previously seen that for an alphabet Σ, Σ^* is defined to be the set of all strings over Σ. If you think of Σ as being a set of length-1 strings, and take its Kleene closure, the result is once again the set of all strings over Σ, and so the two notions of * coincide.

Example 3.6. Let $\Sigma = \{a, b\}$. Then
$\Sigma^0 = \{\varepsilon\}$
$\Sigma^1 = \{a, b\}$
$\Sigma^2 = \{aa, ab, ba, bb\}$
$\Sigma^3 = \{aaa, aab, aba, abb, baa, bab, bba, bbb\}$
etc, so
$\Sigma^* = \{\varepsilon, a, b, aa, ab, ba, bb, aaa, aab, aba, abb, baa, bab, \ldots\}$.

Exercises

1. Let $S = \{\varepsilon, ab, abab\}$ and $T = \{aa, aba, abba, abbba, \ldots\}$. Find the following:
a) S^2 **b)** S^3 **c)** S^* **d)** ST **e)** TS

2. The ***reverse*** of a language L is defined to be $L^R = \{x^R \mid x \in L\}$. Find S^R and T^R for the S and T in the preceding problem.

3. Give an example of a language L such that $L = L^*$.

3.2 Regular Expressions

Though we have used the term *string* throughout to refer to a sequence of symbols from an alphabet, an alternative term that is frequently used is *word*. The analogy seems fairly obvious: strings are made up of "letters" from an alphabet, just as words are in human languages like English. In English, however, there are no particular rules specifying which sequences of letters can be used to form legal English words—even unlikely combinations like *ghth* and *ckstr* have their place. While some formal languages may simply be random collections of arbitrary strings, more interesting languages are those where the strings in the language all share some common structure: $L_1 = \{x \in \{a, b\}^* \mid n_a(x) = n_b(x)\}$; $L_2 = \{\text{legal Java identifiers}\}$; $L_3 = \{\text{legal C++ programs}\}$. In all of these languages, there are structural rules which determine which sequences of symbols are in the language and which aren't. So despite the terminology of "alphabet" and "word" in formal language theory, the concepts don't necessarily match "alphabet" and "word" for human languages. A better parallel is to think of the *alphabet* in a formal language as corresponding to the *words* in a human language; the *words* in a formal language correspond to the *sentences* in a human language, as there are rules (*grammar rules*) which determine how they can legally be constructed.

One way of describing the grammatical structure of the strings in a language is to use a mathematical formalism called a ***regular expression***. A regular expression is a pattern that "matches" strings that have a particular form. For example, consider the language (over alphabet $\Sigma = \{a, b\}$) $L = \{x \mid x \text{ starts and ends with } a\}$. What is the symbol-by-symbol structure of strings in this language? Well, they start with an a, followed by zero or more a's or b's or both, followed by an a. The regular expression $a \cdot (a \mid b)^* \cdot a$ is a pattern that captures this structure and matches any string in L ($\cdot$ and * have their usual meanings, and $\mid$ designates *or*.[1]) Conversely, consider the regular expression $(a \cdot (a \mid b)^*) \mid ((a \mid b)^* \cdot a)$. This is a pattern that matches any string that either has the form "a followed by zero or more a's or b's or both" (i.e. any string that starts with an a) *or* has the form "zero or more a's or b's or both followed by an a" (i.e. any string that ends with an a). Thus the regular expression *generates* the language of all strings that start or end (or both) in an a: this is the set of strings that match the regular expression.

Here are the formal definitions of a regular expression and the language generated by a regular expression:

Definition 3.2. Let Σ be an alphabet. Then the following patterns are ***regular expressions*** over Σ:

1. Φ and ε are regular expressions;

2. a is a regular expression, for each $a \in \Sigma$;

3. if r_1 and r_2 are regular expressions, then so are $r_1 \mid r_2$, $r_1 \cdot r_2$, r_1^* and (r_1) (and of course, r_2^* and (r_2)). As in concatenation of strings, the $\cdot$ is often left out of the second expression. (Note: the order of precedence of operators, from lowest to highest, is $\mid$, $\cdot$, $*$.)

No other patterns are regular expressions.

Definition 3.3. The ***language generated by a regular expression*** r, denoted $L(r)$, is defined as follows:

1. $L(\Phi) = \emptyset$, i.e. no strings match Φ;

2. $L(\varepsilon) = \{\varepsilon\}$, i.e. ε matches only the empty string;

3. $L(a) = \{a\}$, i.e. a matches only the string a;

[1] Various symbols have been used to represent the "or" operation in regular expressions. Both $+$ and $\cup$ have been used for this purpose. In this book, we use the symbol $\mid$ because it is commonly used in computer implementations of regular expressions.

4. $L(r_1 \mid r_2) = L(r_1) \cup L(r_2)$, i.e. $r_1 \mid r_2$ matches strings that match r_1 or r_2 or both;

5. $L(r_1 r_2) = L(r_1)L(r_2)$, i.e. $r_1 r_2$ matches strings of the form "something that matches r_1 followed by something that matches r_2";

6. $L(r_1^*) = (L(r_1))^*$, i.e. r_1^* matches sequences of 0 or more strings each of which matches r_1.

7. $L((r_1)) = L(r_1)$, i.e. (r_1) matches exactly those strings matched by r_1.

Example 3.7. Let $\Sigma = \{a, b\}$, and consider the regular expression $r = a^*b^*$. What is $L(r)$? Well, $L(a) = \{a\}$ so $L(a^*) = (L(a))^* = \{a\}^*$, and $\{a\}^*$ is the set of all strings of zero or more a's, so $L(a^*) = \{\varepsilon, a, aa, aaa, \ldots\}$. Similarly, $L(b^*) = \{\varepsilon, b, bb, bbb, \ldots\}$. Since $L(a^*b^*) = L(a^*)L(b^*) = \{xy \mid x \in L(a^*) \wedge y \in L(b^*)\}$, we have $L(a^*b^*) = \{\varepsilon, a, b, aa, ab, bb, aaa, aab, abb, bbb, \ldots\}$, which is the set of all strings of the form "zero or more a's followed by zero or more b's".

Example 3.8. Let $\Sigma = \{a, b\}$, and consider the regular expression $r = (a \mid aa \mid aaa)(bb)^*$. Since $L(a) = \{a\}$, $L(aa) = L(a)L(a) = \{aa\}$. Similarly, $L(aaa) = \{aaa\}$ and $L(bb) = \{bb\}$. Now $L(a \mid aa \mid aaa) = L(a) \cup L(aa) \cup L(aaa) = \{a, aa, aaa\}$, and $L((bb)^*) = (L((bb)))^* = (L(bb))^*$ (the last equality is from clause 7 of Definition 3.3), and $(L(bb))^* = \{bb\}^* = \{\varepsilon, bb, bbbb, \ldots\}$. So $L(r)$ is the set of strings formed by concatenating a or aa or aaa with zero or more pairs of b's.

Definition 3.4. A language is *regular* if it is generated by a regular expression.

Clearly the union of two regular languages is regular; likewise, the concatenation of regular languages is regular; and the Kleene closure of a regular language is regular. It is less clear whether the intersection of regular languages is always regular; nor is it clear whether the complement of a regular language is guaranteed to be regular. These are questions that will be taken up in Section 3.6.

Regular languages, then, are languages whose strings' structure can be described in a very formal, mathematical way. The fact that a language can be "mechanically" described or generated means that we are likely to be able to get a computer to recognize strings in that language. We will pursue the question of mechanical language recognition in Section 3.4, and subsequently will see that our first attempt to model mechanical language recognition does in fact produce a family of "machines" that recognize exactly the regular languages. But first, in the next section, we will look at some practical applications of regular expressions.

Exercises

1. Give English-language descriptions of the languages generated by the following regular expressions.
a) $(a \mid b)^*$ **b)** $a^* \mid b^*$ **c)** $b^*(ab^*ab^*)^*$ **d)** $b^*(abb^*)$

2. Give regular expressions over $\Sigma = \{a, b\}$ that generate the following languages.
a) $L_1 = \{x \mid x \text{ contains 3 consecutive } a\text{'s}\}$
b) $L_2 = \{x \mid x \text{ has even length}\}$
c) $L_3 = \{x \mid n_b(x) = 2 \bmod 3\}$
d) $L_4 = \{x \mid x \text{ contains the substring } aaba\}$
e) $L_5 = \{x \mid n_b(x) < 2\}$
f) $L_6 = \{x \mid x \text{ doesn't end in } aa\}$

3. Prove that all finite languages are regular.

3.3 Application: Using Regular Expressions

A common operation when editing text is to search for a given string of characters, sometimes with the purpose of replacing it with another string. Many "search and replace" facilities have the option of using regular expressions instead of simple strings of characters. A regular expression describes a language, that is, a *set* of strings. We can think of a regular expression as a ***pattern*** that matches certain strings, namely all the strings in the language described by the regular expression. When a regular expression is used in a search operation, the goal is to find a string that matches the expression. This type of ***pattern matching*** is very useful.

The ability to do pattern matching with regular expressions is provided in many text editors, including *jedit* and *kwrite*. Programming languages often come with libraries for working with regular expressions. Java (as of version 1.4) provides regular expression handling though a package named *java.util.regexp*. C++ typically provides a header file named *regexp.h* for the same purpose. In all these applications, many new notations are added to the syntax to make it more convenient to use. The syntax can vary from one implementation to another, but most implementations include the capabilities discussed in this section.

In applications of regular expressions, the alphabet usually includes all the characters on the keyboard. This leads to a problem, because regular expressions actually use two types of symbols: symbols that are members of the alphabet and special symbols such a "`*`" and "`)`" that are used to construct expressions. These special symbols, which are not part of the language being described but are used in the description, are called ***meta-characters***. The problem is, when the alphabet includes all the available characters, what do we do about meta-characters? If the language that we are describing uses the "`*`" character, for example, how can we represent the Kleene star operation?

The solution is to use a so-called "escape character," which is usually the backslash, `\`. We agree, for example, that the notation `\*` refers to the symbol `*` that is a member of the alphabet, while `*` by itself is the meta-character that represents the Kleene star operation. Similarly, `(` and `)` are the meta-characters that are used for grouping, while the corresponding characters in the language are written as `\(` and `\)`. For example, a regular expression that matches the string `a*b` repeated any number of times would be written: `(a\*b)*`. The backslash is also used to represent certain non-printing characters. For example, a tab is represented as `\t` and a new line character is `\n`.

We introduce two new common operations on regular expressions and two new meta-characters to represent them. The first operation is represented by the meta-character `+`: If `r` is a regular expression, then `r+` represents the occurrence of `r` one or more times. The second operation is represented by `?`: The notation `r?` represents an occurrence of `r` zero or one times. That is to say, `r?` represents an optional occurrence of `r`. Note that these operations are introduced for convenience only and do not represent any real increase in the power. In fact, `r+` is exactly equivalent to `rr*`, and `r?` is equivalent to `(r|`ε`)` (except that in applications there is generally no equivalent to ε).

To make it easier to deal with the large number of characters in the alphabet, ***character classes*** are introduced. A character class consists of a list of characters enclosed between brackets, `[` and `]`. (The brackets are meta-characters.) A character class matches a single character, which can be any of the characters in the list. For example, `[0123456789]` matches any one of the digits 0 through 9. The same thing could be expressed as `(0|1|2|3|4|5|6|7|8|9)`, so once again we have added only convenience, not new representational power. For even more convenience, a hyphen can be included in a character class to indicate a range of characters. This means that `[0123456789]` could also be written as `[0-9]` and that the regular expression `[a-z]` will match any single lowercase letter. A character class can include multiple ranges, so that `[a-zA-Z]` will match any letter, lower- or uppercase. The period (`.`) is a meta-character that will match any single character, except (in most implementations) for an end-of-line. These notations can, of course, be used in more complex regular expressions. For example, `[A-Z][a-zA-Z]*` will match any capitalized word, and `\(.*\)` matches any string of characters enclosed in parentheses.

In most implementations, the meta-character `^` can be used in a regular expression to match the beginning of a line of text, so that the expression `^[a-zA-Z]+` will only match a word that occurs at the start of a line. Similarly, `$` is used as a meta-character to match the end of a line. Some implementations also have a way of matching beginnings and ends of words. Typically, `\b` will match such "word boundaries." Using this notation, the pattern `\band\b` will match the string "and" when it occurs as a word, but will not match the a-n-d in the word "random." We are going a bit beyond basic regular expressions here: Previously, we only thought of a regular expression as something that either will match or will not match a given string in its entirety. When we use a regular expression for a search operation, however, we want to find a *substring* of a given string that matches the expression. The notations `^`, `$` and `\b` put a restrictions on *where* the matching substring can be located in the string.

When regular expressions are used in search-and-replace operations, a regular expression is used for the search pattern. A search is made in a (typically long) string for a substring that matches the pattern, and then the substring is replaced by a specified replacement pattern. The replacement pattern is not used for matching and is not a regular expression. However, it can be more than just a simple string. It's possible to include parts of the substring that is being replaced in the replacement string. The notations `\0`, `\1`, ..., `\9` are used for this purpose. The first of these, `\0`, stands for the entire substring that is being replaced. The others are only available when parentheses are used in the search pattern. The notation `\1` stands for "the part of the substring that matched the part of the search pattern beginning with the first `(` in the pattern and ending with the matching `)`." Similarly, `\2` represents whatever matched the part of the search pattern between the second pair of parentheses, and so on.

Suppose, for example, that you would like to search for a name in the form *last-name, first-name* and replace it with the same name in the form *first-name last-name*. For example, "Reeves, Keanu" should be converted to "Keanu Reeves". Assuming that names contain only letters, this could be done using the search pattern `([A-Za-z]+), ([A-Za-z]+)` and the replacement pattern `\2 \1`. When the match is made, the first `([A-Za-z]+)` will match "Reeves," so that in the replacement pattern, `\1` represents the substring "Reeves". Similarly, `\2` will represent "Keanu". Note that the parentheses are included in the search pattern *only* to specify what parts of the string are represented by `\1` and `\2`. In practice, you might use

`^([A-Za-z]+), ([A-Za-z])$` as the search pattern to constrain it so that it will only match a complete line of text. By using a "global" search-and-replace, you could convert an entire file of names from one format to the other in a single operation.

Regular expressions are a powerful and useful technique that should be part of any computer scientist's toolbox. This section has given you a taste of what they can do, but you should check out the specific capabilities of the regular expression implementation in the tools and programming languages that you use.

Exercises

1. The backslash is itself a meta-character. Suppose that you want to match a string that contains a backslash character. How do you suppose you would represent the backslash in the regular expression?

2. Using the notation introduced in this section, write a regular expression that could be used to match each of the following:

a) Any sequence of letters (upper- or lowercase) that includes the letter Z (in uppercase).

b) Any eleven-digit telephone number written in the form `(xxx)xxx-xxxx`.

c) Any eleven-digit telephone number *either* in the form `(xxx)xxx-xxxx` or `xxx-xxx-xxxx`.

d) A non-negative real number with an optional decimal part. The expression should match numbers such as 17, 183.9999, 182., 0, 0.001, and 21333.2.

e) A complete line of text that contains only letters.

f) A C++ style one-line comment consisting of `//` and all the following characters up to the end-of-line.

3. Give a search pattern and a replace pattern that could be used to perform the following conversions:

a) Convert a string that is enclosed in a pair of double quotes to the same string with the double quotes replaced by single quotes.

b) Convert seven-digit telephone numbers in the format `xxx-xxx-xxxx` to the format `(xxx)xxx-xxxx`.

c) Convert C++ one-line comments, consisting of characters between `//` and end-of-line, to C style comments enclosed between `/*` and `*/`.

d) Convert any number of consecutive spaces and tabs to a single space.

4. In some implementations of "regular expressions," the notations `\1`, `\2`, and so on can occur in a search pattern. For example, consider the search pattern `^([a-zA-Z]).*\1$`. Here, `\1` represents a recurrence of the same substring that matched `[a-zA-Z]`, the part of the pattern between the first pair of parentheses. The entire pattern, therefore, will match a line of text that begins and ends with the same letter. Using this notation, write a pattern that matches all strings in the language $L = \{a^n b a^n \mid n \geq 0\}$. (Later in this chapter, we will see that L is *not* a regular language, so allowing the use of \1 in a "regular expression" means that it's not really a regular expression at all! This notation can add a real increase in expressive power to the patterns that contain it.)

3.4 Finite-State Automata

We have seen how regular expressions can be used to generate languages mechanically. How might languages be recognized mechanically? The question is of interest because if we can mechanically recognize languages like $L = \{$all legal C++ programs that will not go into infinite loops on any input$\}$, then it would be possible to write über-compilers that can do semantic error-checking like testing for infinite loops, in addition to the syntactic error-checking they currently do.

What formalism might we use to model what it means to recognize a language "mechanically"? We look for inspiration to a language-recognizer with which we are all familiar, and which we've already in fact mentioned: a compiler. Consider how a C++ compiler might handle recognizing a legal *if* statement. Having seen the word *if*, the compiler will be in a *state* or *phase of its execution* where it expects to see a '('; in this state, any other character will put the compiler in a "failure" state. If the compiler does in fact see a '(' next, it will then be in an "expecting a boolean condition" state; if it sees a sequence of symbols that make up a legal boolean condition, it will then be in an "expecting a ')'" state; and then "expecting a '{' or a legal statement"; and so on. Thus one can think of the compiler as being in a series of states; on seeing a new input symbol, it moves on to a new state; and this sequence of transitions eventually leads to either a "failure" state (if the *if* statement is not syntactically correct) or a "success" state (if the *if* statement is legal). We isolate these three concepts—states, input-inspired transitions from state to state, and "accepting" vs "non-accepting" states—as the key features of a mechanical language-recognizer, and capture them in a model called a *finite-state automaton.* (Whether this is a successful distillation of the essence of mechanical language recognition remains to be seen; the question will be taken up later in this chapter.)

A ***finite-state automaton (FSA)***, then, is a machine which takes, as input, a finite string of symbols from some alphabet Σ. There is a finite set of ***states*** in which the machine can find itself. The state it is in before consuming any input is called the ***start state***. Some of the states are ***accepting*** or ***final***. If the machine ends in such a state after completely consuming an input string, the string is said to be ***accepted*** by the machine. The actual functioning of the machine is described by something called a ***transition function***, which specifies what happens if the machine is in a particular state and looking at a particular input symbol. ("What happens" means "in which state does the machine end up".)

Example 3.9. Below is a table that describes the transition function of a finite-state automaton with states p, q, and r, on inputs 0 and 1.

	p	q	r
0	p	q	r
1	q	r	r

The table indicates, for example, that if the FSA were in state p and consumed a 1, it would move to state q.

FSAs actually come in two flavours depending on what properties you require of the transition function. We will look first at a class of FSAs called deterministic finite-state automata (DFAs). In these machines, the current state of the machine and the current input symbol together determine exactly which state the machine ends up in: for every <current state, current input symbol> pair, there is exactly one possible next state for the machine.

Definition 3.5. Formally, a ***deterministic finite-state automaton*** M is specified by 5 components: $M = (Q, \Sigma, q_0, \delta, F)$ where

- Q is a finite set of states;
- Σ is an alphabet called the *input alphabet*;
- $q_0 \in Q$ is a state which is designated as the *start state*;
- F is a subset of Q; the states in F are states designated as *final* or *accepting* states;
- δ is a transition function that takes <state, input symbol> pairs and maps each one to a state: $\delta : Q \times \Sigma \rightarrow Q$. To say $\delta(q, a) = q'$ means that if the machine is in state q and the input symbol a is consumed, then the machine will move into state q'. The function δ must be a total function, meaning that $\delta(q, a)$ must be defined for every state q and every input symbol a. (Recall also that, according to the definition of a function, there can be only one output for any particular input. This means that for any given q and a, $\delta(q, a)$ can

have only one value. This is what makes the finite-state automaton deterministic: given the current state and input symbol, there is only one possible move the machine can make.)

Example 3.10. The transition function described by the table in the preceding example is that of a DFA. If we take p to be the start state and r to be a final state, then the formal description of the resulting machine is $M = (\{p, q, r\}, \{0, 1\}, p, \delta, \{r\})$, where δ is given by

$$\begin{array}{ll} \delta(p,0) = p & \delta(p,1) = q \\ \delta(q,0) = q & \delta(q,1) = r \\ \delta(r,0) = r & \delta(r,1) = r \end{array}$$

The transition function δ describes only individual steps of the machine as individual input symbols are consumed. However, we will often want to refer to "the state the automaton will be in if it starts in state q and consumes input string w", where w is a string of input symbols rather than a single symbol. Following the usual practice of using * to designate "0 or more", we define $\delta^*(q, w)$ as a convenient shorthand for "the state that the automaton will be in if it starts in state q and consumes the input string w". For any string, it is easy to see, based on δ, what steps the machine will make as those symbols are consumed, and what $\delta^*(q, w)$ will be for any q and w. Note that if no input is consumed, a DFA makes no move, and so $\delta^*(q, \varepsilon) = q$ for any state q.[2]

Example 3.11. Let M be the automaton in the preceding example. Then, for example:

$\delta^*(p, 001) = q$, since $\delta(p, 0) = p$, $\delta(p, 0) = p$, and $\delta(p, 1) = q$;
$\delta^*(p, 01000) = q$;
$\delta^*(p, 1111) = r$;
$\delta^*(q, 0010) = r$.

We have divided the states of a DFA into accepting and non-accepting states, with the idea that some strings will be recognized as "legal" by the automaton, and some not. Formally:

Definition 3.6. Let $M = (Q, \Sigma, q_0, \delta, F)$. A string $w \in \Sigma^*$ is ***accepted*** by M iff $\delta^*(q_0, w) \in F$. (Don't get confused by the notation. Remember, it's just a shorter and neater way of saying "$w \in \Sigma^*$ is accepted by M if

[2] δ^* can be defined formally by saying that $\delta^*(q, \varepsilon) = q$ for every state q, and $\delta^*(q, ax) = \delta^*(\delta(q, a), x)$ for any state q, $a \in \Sigma$ and $x \in \Sigma^*$. Note that this is a recursive definition.

and only if the state that M will end up in if it starts in q_0 and consumes w is one of the states in F.")

The ***language accepted by*** M, denoted $L(M)$, is the set of all strings $w \in \Sigma^*$ that are accepted by M: $L(M) = \{w \in \Sigma^* \mid \delta^*(q_0, w) \in F\}$.

Note that we sometimes use a slightly different phrasing and say that a language L is accepted by some machine M. We don't mean by this that L *and maybe some other strings* are accepted by M; we mean $L = L(M)$, i.e. L is *exactly* the set of strings accepted by M.

It may not be easy, looking at a formal specification of a DFA, to determine what language that automaton accepts. Fortunately, the mathematical description of the automaton $M = (Q, \Sigma, q_0, \delta, F)$ can be neatly and helpfully captured in a picture called a ***transition diagram***. Consider again the DFA of the two preceding examples. It can be represented pictorially as:

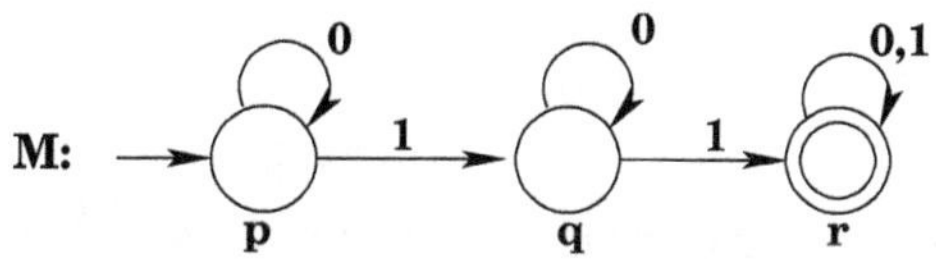

The arrow on the left indicates that p is the start state; double circles indicate that a state is accepting. Looking at this picture, it should be fairly easy to see that the language accepted by the DFA M is $L(M) = \{x \in \{0,1\}^* \mid n_1(x) \geq 2\}$.

Example 3.12. Find the language accepted by the DFA shown below (and describe it using a regular expression!)

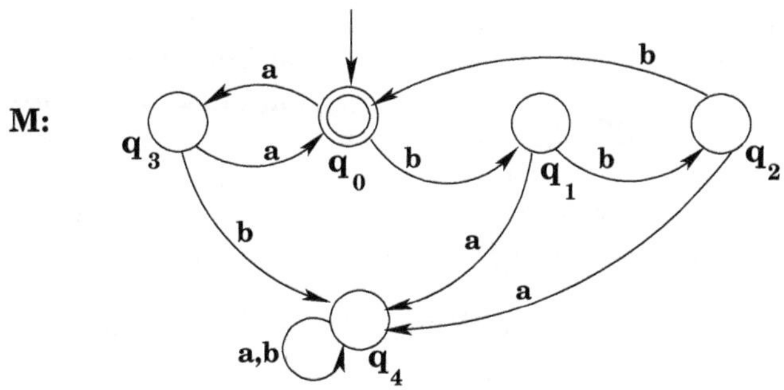

The start state of M is accepting, which means $\varepsilon \in L(M)$. If M is in state q_0, a sequence of two a's or three b's will move M back to q_0 and hence be accepted. So $L(M) = L((aa \mid bbb)^*)$.

The state q_4 in the preceding example is often called a *garbage* or *trap* state: it is a non-accepting state which, once reached by the machine, cannot be escaped. It is fairly common to omit such states from transition diagrams. For example, one is likely to see the diagram:

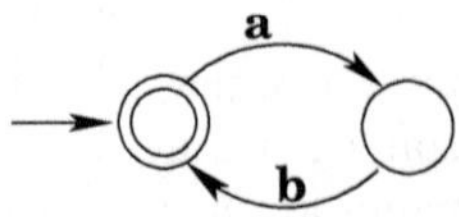

Note that this cannot be a complete DFA, because a DFA is required to have a transition defined for every state-input pair. The diagram is "short for" the full diagram:

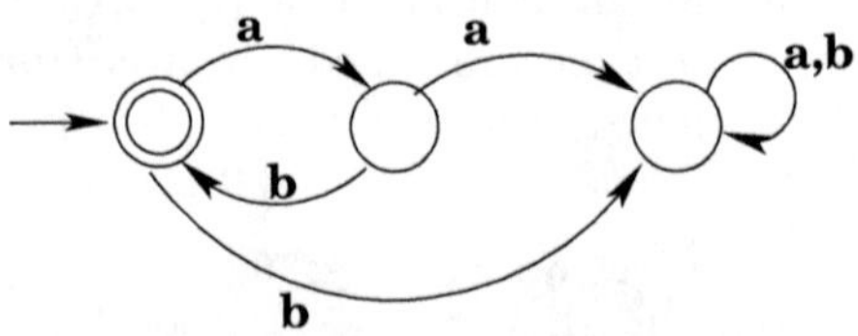

As well as recognizing what language is accepted by a given DFA, we often want to do the reverse and come up with a DFA that accepts a given language. Building DFAs for specified languages is an art, not a science. There is no algorithm that you can apply to produce a DFA from an English-language description of the set of strings the DFA should accept. On the other hand, it is not generally successful, either, to simply write down a half-dozen strings that are in the language and design a DFA to accept those strings—invariably there are strings that are in the language that aren't accepted, and other strings that aren't in the language that are accepted. So how do you go about building DFAs that accept all and only the strings they're supposed to accept? The best advice I can give is to think about relevant characteristics that determine whether a string is in the language or not, and to think about what the possible values or "states" of those characteristics are; then build a machine that has a state corresponding to each possible combination of values of relevant characteristics, and determine how the consumption of inputs affects those values. I'll illustrate what I mean with a couple of examples.

Example 3.13. Find a DFA with input alphabet $\Sigma = \{a, b\}$ that accepts the language $L = \{w \in \Sigma^* \mid n_a(w)$ and $n_b(w)$ are both even $\}$.

The characteristics that determine whether or not a string w is in L are the parity of $n_a(w)$ and $n_b(w)$. There are four possible combinations of "values" for these characteristics: both numbers could be even, both could be odd, the first could be odd and the second even, or the first could be even and the second odd. So we build a machine with four states q_1, q_2, q_3, q_4 corresponding to the four cases. We want to set up δ so that the machine will be in state q_1 exactly when it has consumed a string with an even number of a's and an even number of b's, in state q_2 exactly when it has consumed a string with an odd number of a's and an odd number of b's, and so on.

To do this, we first make the state q_1 into our start state, because the DFA will be in the start state after consuming the empty string ε, and ε has an even number (zero) of both a's and b's. Now we add transitions by reasoning about how the parity of a's and b's is changed by additional input. For instance, if the machine is in q_1 (meaning an even number of a's and an even number of b's have been seen) and a further a is consumed, then we want the machine to move to state q_3, since the machine has now consumed an odd number of a's and still an even number of b's. So we add the transition $\delta(q_1, a) = q_3$ to the machine. Similarly, if the machine is in q_2 (meaning an odd number of a's and an odd number of b's have been seen) and a further b is consumed, then we want the machine to move to state q_3 again, since the machine has still consumed an odd number of a's, and now an even number of b's. So we add the transition $\delta(q_2, b) = q_3$ to the machine. Similar reasoning produces a total of eight transitions, one for each state-input pair. Finally, we have to decide which states should be final states. The only state that corresponds to the desired criteria for the language L is q_1, so we make q_1 a final state. The complete machine is shown below.

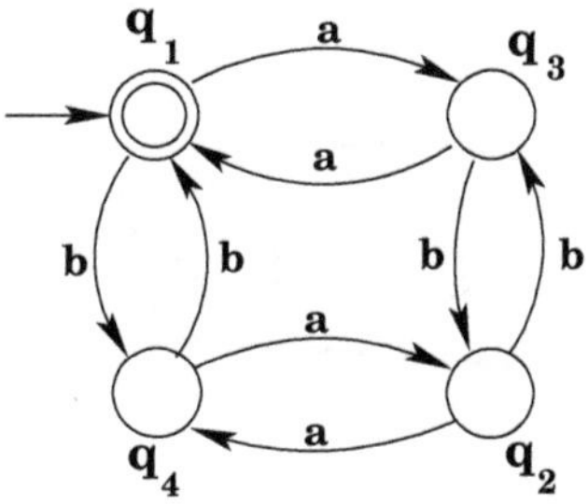

Example 3.14. Find a DFA with input alphabet $\Sigma = \{a, b\}$ that accepts the language $L = \{w \in \Sigma^* \mid n_a(w) \text{ is divisible by } 3\}$.

The relevant characteristic here is of course whether or not the number of a's in a string is divisible by 3, perhaps suggesting a two-state machine. But in fact, there is more than one way for a number to not be divisible by 3: dividing the number by 3 could produce a remainder of either 1 or 2 (a remainder of 0 corresponds to the number in fact being divisible by 3).

So we build a machine with three states q_0, q_1, q_2, and add transitions so that the machine will be in state q_0 exactly when the number of a's it has consumed is evenly divisible by 3, in state q_1 exactly when the number of a's it has consumed is equivalent to 1 mod 3, and similarly for q_2. State q_0 will be the start state, as ε has 0 a's and 0 is divisible by 3. The completed machine is shown below. Notice that because the consumption of a b does not affect the only relevant characteristic, b's do not cause changes of state.

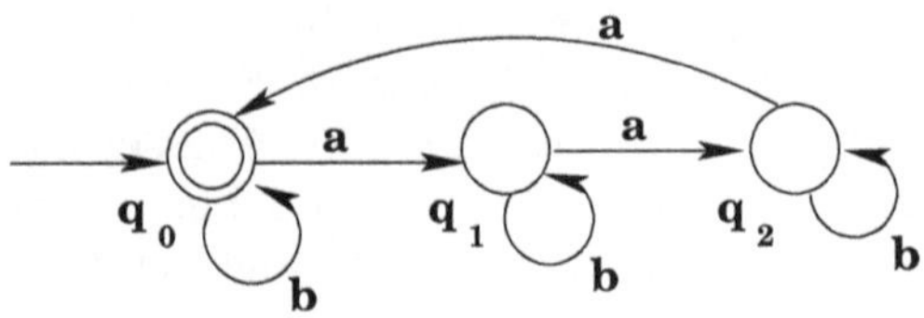

Example 3.15. Find a DFA with input alphabet $\Sigma = \{a, b\}$ that accepts the language $L = \{w \in \Sigma^* \mid w \text{ contains three consecutive a's }\}$.

Again, it is not quite so simple as making a two-state machine where the states correspond to "have seen aaa" and "have not seen aaa". Think dynamically: as you move through the input string, how do you arrive at the goal of having seen three consecutive a's? You might have seen two consecutive a's and still need a third, or you might just have seen one a and be looking for two more to come immediately, or you might just have seen a b and be right back at the beginning as far as seeing 3 consecutive a's goes. So once again there will be three states, with the "last symbol was not an a" state being the start state. The complete automaton is shown below.

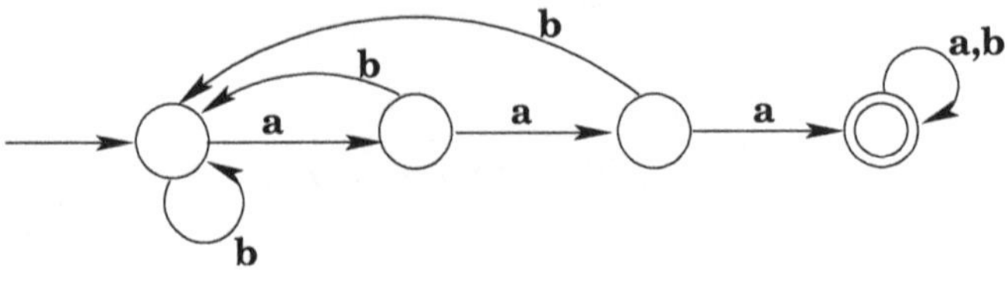

Exercises

1. Give DFAs that accept the following languages over $\Sigma = \{a, b\}$.
   a) $L_1 = \{x \mid x \text{ contains the substring } aba\}$
   b) $L_2 = L(a^*b^*)$
   c) $L_3 = \{x \mid n_a(x) + n_b(x) \text{ is even }\}$
   d) $L_4 = \{x \mid n_a(x) \text{ is a multiple of 5 }\}$
   e) $L_5 = \{x \mid x \text{ does not contain the substring } abb\}$

f) $L_6 = \{x \mid x \text{ has no } a\text{'s in the even positions}\}$
g) $L_7 = L(aa^* \mid aba^*b^*)$

2. What languages do the following DFAs accept?

a)

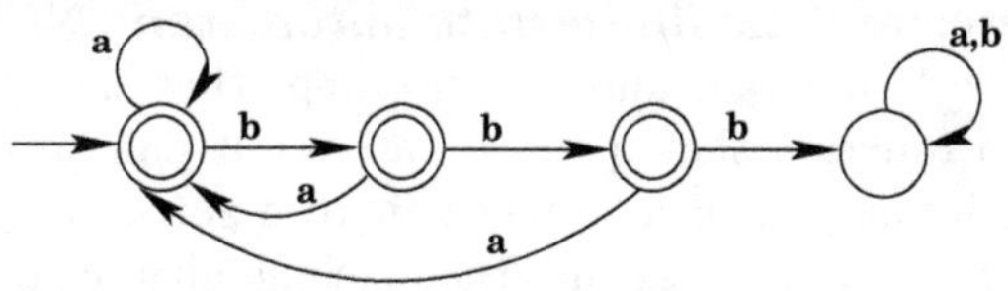

b)

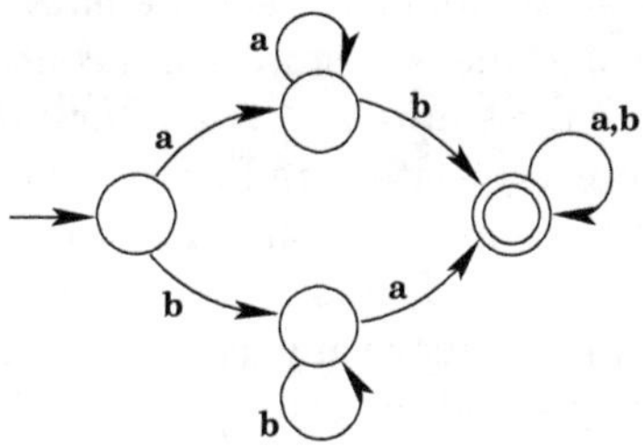

3. Let $\Sigma = \{0, 1\}$. Give a DFA that accepts the language

$$L = \{x \in \Sigma^* \mid x \text{ is the binary representation of an integer divisible by } 3\}.$$

3.5 Nondeterministic Finite-State Automata

As mentioned briefly above, there is an alternative school of though as to what properties should be required of a finite-state automaton's transition function. Recall our motivating example of a C++ compiler and a legal *if* statement. In our description, we had the compiler in an "expecting a ')'" state; on seeing a ')', the compiler moved into an "expecting a '{' or a legal statement" state. An alternative way to view this would be to say that the compiler, on seeing a ')', could move into one of two different states: it could move to an "expecting a '{'" state **or** move to an "expecting a legal statement" state. Thus, from a single state, on input ')', the compiler has multiple moves. This alternative interpretation is not allowed by the DFA model. A second point on which one might question the DFA model is the fact that input must be consumed for the machine to change state. Think of the syntax for C++ function declarations. The return type of a function need not be specified (the default is taken to be *int*). The start state of the compiler when parsing a function declaration might be "expecting a return type"; then with no input consumed, the compiler can move to the state "expecting a legal function name". To model this, it might seem reasonable to allow transitions that do not require the consumption of input (such transitions are called ***ε-transitions***). Again, this is not supported by the DFA abstraction. There is, therefore, a second class of finite-state automata that people study, the class of nondeterministic finite-state automata.

A ***nondeterministic finite-state automaton (NFA)*** is the same as a deterministic finite-state automaton except that the transition function is no longer a function that maps a state-input pair to a state; rather, it maps a state-input pair or a state-ε pair to a **set** of states. No longer do we have $\delta(q, a) = q'$, meaning that the machine must change to state q' if it is in state q and consumes an a. Rather, we have $\partial(q, a) = \{q_1, q_2, \ldots, q_n\}$, meaning that if the machine is in state q and consumes an a, it might move directly to any one of the states $q_1, \ldots, q_n$. Note that the set of next states $\partial(q, a)$ is defined for every state q and every input symbol a, but for some q's and a's it could be empty, or contain just one state (there don't **have** to be multiple next states). The function ∂ must also specify whether it is possible for the machine to make any moves without input being consumed, i.e. $\partial(q, \varepsilon)$ must be specified for every state q. Again, it is quite possible that $\partial(q, \varepsilon)$ may be empty for some states q: there need not be ε-transitions out of q.

Definition 3.7. Formally, a nondeterministic finite-state automaton M is specified by 5 components: $M = (Q, \Sigma, q_0, \partial, F)$ where

- Q, Σ, q_0 and F are as in the definition of DFAs;
- ∂ is a transition function that takes <state, input symbol> pairs and maps each one to a set of states. To say $\partial(q, a) = \{q_1, q_2, \ldots, q_n\}$ means that if the machine is in state q and the input symbol a is consumed, then the machine may move directly into any one of states $q_1, q_2, \ldots, q_n$. The function ∂ must also be defined for every <state,ε> pair. To say $\partial(q, \varepsilon) = \{q_1, q_2, \ldots, q_n\}$ means that there are direct ε-transitions from state q to each of $q_1, q_2, \ldots, q_n$.

  The formal description of the function ∂ is $\partial : Q \times (\Sigma \cup \{\varepsilon\}) \to \mathcal{P}(Q)$.

The function ∂ describes how the machine functions on zero or one input symbol. As with DFAs, we will often want to refer to the behavior of the machine on a string of inputs, and so we use the notation $\partial^*(q, w)$ as shorthand for "the set of states in which the machine might be if it starts in state q and consumes input string w". As with DFAs, $\partial^*(q, w)$ is determined by the specification of ∂. Note that for every state q, $\partial^*(q, \varepsilon)$ contains at least q, and may contain additional states if there are (sequences of) ε-transitions out of q.

We do have to think a bit carefully about what it means for an NFA to accept a string w. Suppose $\partial^*(q_0, w)$ contains both accepting and non-accepting states, i.e. the machine could end in an accepting state after consuming w, but it might also end in a non-accepting state. Should we consider the machine to accept w, or should we require every state in $\partial^*(q_0, w)$ to be accepting before we admit w to the ranks of the accepted? Think of the C++ compiler again: provided that an *if* statement fits one of the legal syntax specifications, the compiler will accept it. So we take as the definition of acceptance by an NFA: A string w is accepted by an NFA provided that at least one of the states in $\partial^*(q_0, w)$ is an accepting state. That is, if there is some sequence of steps of the machine that consumes w and leaves the machine in an accepting state, then the machine accepts w. Formally:

Definition 3.8. Let $M = (Q, \Sigma, q_0, \partial, F)$ be a nondeterministic finite-state automaton. The string $w \in \Sigma^*$ is ***accepted*** by M iff $\partial^*(q_0, w)$ contains at least one state $q_F \in F$.

The ***language accepted by*** M, denoted $L(M)$, is the set of all strings $w \in \Sigma^*$ that are accepted by M: $L(M) = \{w \in \Sigma^* \mid \partial^*(q_0, w) \cap F \neq \emptyset\}$.

Example 3.16. The NFA shown below accepts all strings of a's and b's in which the second-to-last symbol is a.

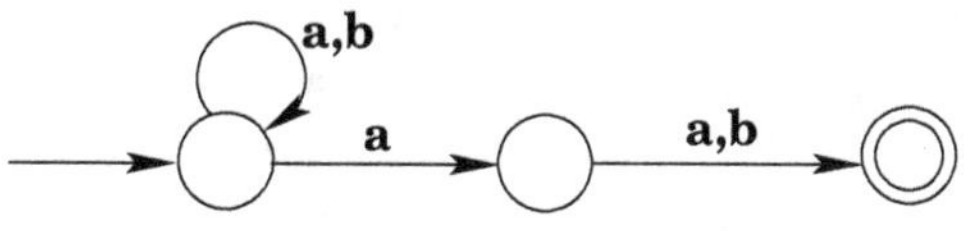

It should be fairly clear that every language that is accepted by a DFA is also accepted by an NFA. Pictorially, a DFA looks exactly like an NFA (an NFA that doesn't happen to have any ε-transitions or multiple same-label transitions from any state), though there is slightly more going on behind the scenes. Formally, given the DFA $M = (Q, \Sigma, q_0, \delta, F)$, you can build an NFA $M' = (Q, \Sigma, q_0, \partial, F)$ where 4 of the 5 components are the same and where every transition $\delta(q, a) = q'$ has been replaced by $\partial(q, a) = \{q'\}$.

But is the reverse true? Can any NFA-recognized language be recognized by a DFA? Look, for example, at the language in Example 3.16. Can you come up with a DFA that accepts this language? Try it. It's pretty difficult to do. But does that mean that there really is **no** DFA that accepts the language, or only that we haven't been clever enough to find one?

It turns out that the limitation is in fact in our cleverness, and not in the power of DFAs.

Theorem 3.2. *Every language that is accepted by an NFA is accepted by a DFA.*

Proof. Suppose we are given an NFA $N = (P, \Sigma, p_0, \partial, F_p)$, and we want to build a DFA $D = (Q, \Sigma, q_0, \delta, F_q)$ that accepts the same language. The idea is to make the states in D correspond to *subsets* of N's states, and then to set up D's transition function δ so that for any string w, $\delta^*(q_0, w)$ corresponds to $\partial^*(p_0, w)$; i.e. the **single** state that w gets you to in D corresponds to the **set** of states that w could get you to in N. If any of those states is accepting in N, w would be accepted by N, and so the corresponding state in D would be made accepting as well.

So how do we make this work? The first thing to do is to deal with a start state q_0 for D. If we're going to make this state correspond to a subset of N's states, what subset should it be? Well, remember (1) that in any DFA, $\delta^*(q_0, \varepsilon) = q_0$; and (2) we want to make $\delta^*(q_0, w)$ correspond to $\partial^*(p_0, w)$ for every w. Putting these two limitations together tells us that we should make q_0 correspond to $\partial^*(p_0, \varepsilon)$. So q_0 corresponds to the subset of all of N's states that can be reached with no input.

Now we progressively set up D's transition function δ by repeatedly doing the following:

– find a state q that has been added to D but whose out-transitions have not yet been added. (Note that q_0 initially fits this description.) Remember that the state q corresponds to some subset $\{p_1, \ldots, p_n\}$ of N's states.

– for each input symbol a, look at all N's states that can be reached from any one of $p_1, \ldots, p_n$ by consuming a (perhaps making some ε-transitions as well). That is, look at $\partial^*(p_1, a) \cup \ldots \cup \partial^*(p_n, a)$. If there is not already a DFA state q' that corresponds to this subset of N's states, then add one, and add the transition $\delta(q, a) = q'$ to D's transitions.

The above process must halt eventually, as there are only a finite number of states n in the NFA, and therefore there can be at most 2^n states in the DFA, as that is the number of subsets of the NFA's states. The final states of the new DFA are those where at least one of the associated NFA states is an accepting state of the NFA.

Can we now argue that $L(D) = L(N)$? We can, if we can argue that $\delta^*(q_0, w)$ corresponds to $\partial^*(p_0, w)$ for all $w \in \Sigma^*$: if this latter property holds, then $w \in L(D)$ iff $\delta^*(q_0, w)$ is accepting, which we made be so iff $\partial^*(p_0, w)$ contains an accepting state of N, which happens iff N accepts w i.e. iff $w \in L(N)$.

So can we argue that $\delta^*(q_0, w)$ does in fact correspond to $\partial^*(p_0, w)$ for all w? We can, using induction on the length of w.

First, a preliminary observation. Suppose $w = xa$, i.e. w is the string x followed by the single symbol a. How are $\partial^*(p_0, x)$ and $\partial^*(p_0, w)$ related?

Well, recall that $\partial^*(p_0, x)$ is the set of all states that N can reach when it starts in p_0 and consumes x: $\partial^*(p_0, x) = \{p_1, \ldots, p_n\}$ for some states $p_1, \ldots, p_n$. Now, w is just x with an additional a, so where might N end up if it starts in p_0 and consumes w? We know that x gets N to p_1 or ... or p_n, so xa gets N to any state that can be reached from p_1 with an a (and maybe some ε-transitions), and to any state that can be reached from p_2 with an a (and maybe some ε-transitions), etc. Thus, our relationship between $\partial^*(p_0, x)$ and $\partial^*(p_0, w)$ is that if $\partial^*(p_0, x) = \{p_1, \ldots, p_n\}$, then $\partial^*(p_0, w) = \partial^*(p_1, a) \cup \ldots \cup \partial^*(p_n, a)$. With this observation in hand, let's proceed to our proof by induction.

We want to prove that $\delta^*(q_0, w)$ corresponds to $\partial^*(p_0, w)$ for all $w \in \Sigma^*$. We use induction on the length of w.

1. Base case: Suppose w has length 0. The only string w with length 0 is ε, so we want to show that $\delta^*(q_0, \varepsilon)$ corresponds to $\partial^*(p_0, \varepsilon)$. Well, $\delta^*(q_0, \varepsilon) = q_0$, since in a DFA, $\delta^*(q, \varepsilon) = q$ for any state q. We explicitly made q_0 correspond to $\partial^*(p_0, \varepsilon)$, and so the property holds for w with length 0.

2. Inductive case: Assume that the desired property holds for some number n, i.e. that $\delta^*(q_0, x)$ corresponds to $\partial^*(p_0, x)$ for all x with length n. Look at an arbitrary string w with length $n + 1$. We want to show that $\delta^*(q_0, w)$ corresponds to $\partial^*(p_0, w)$. Well, the string w must look like xa for some string x (whose length is n) and some symbol a. By our inductive hypothesis, we know $\delta^*(q_0, x)$ corresponds to $\partial^*(p_0, x)$. We know $\partial^*(p_0, x)$ is a set of N's states, say $\partial^*(p_0, x) = \{p_1, \ldots, p_n\}$.

   At this point, our subsequent reasoning might be a bit clearer if we give explicit names to $\delta^*(q_0, w)$ (the state D reaches on input w) and $\delta^*(q_0, x)$ (the state D reaches on input x). Call $\delta^*(q_0, w)$ q_w, and call $\delta^*(q_0, x)$ q_x. We know, because $w = xa$, there must be an a-transition from q_x to q_w. Look at how we added transitions to δ: the fact that there is an a-transition from q_x to q_w means that q_w corresponds to the set $\partial^*(p_1, a) \cup \ldots \cup \partial^*(p_n, a)$ of N's states. By our preliminary observation, $\partial^*(p_1, a) \cup \ldots \cup \partial^*(p_n, a)$ is just $\partial^*(p_0, w)$. So q_w (or $\delta^*(q_0, w)$) corresponds to $\partial^*(p_0, w)$, which is what we wanted to prove. Since w was an arbitrary string of length $n + 1$, we have shown that the property holds for $n + 1$.

Altogether, we have shown by induction that $\delta^*(q_0, w)$ corresponds to $\partial^*(p_0, w)$ for all $w \in \Sigma^*$. As indicated at the very beginning of this proof, that is enough to prove that $L(D) = L(N)$. So for any NFA N, we can find a DFA D that accepts the same language. □

Example 3.17. Consider the NFA shown below.

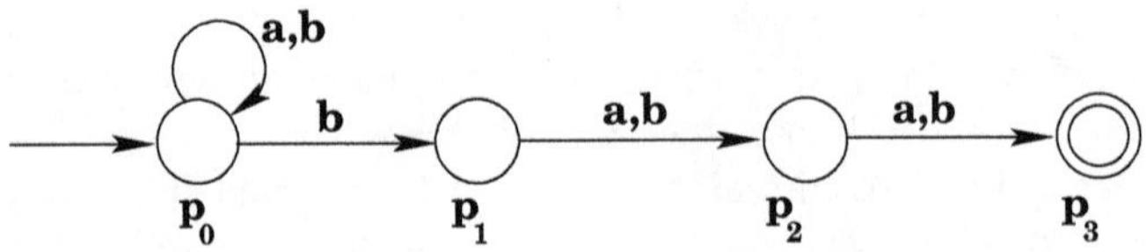

We start by looking at $\partial^*(p_0, \varepsilon)$, and then add transitions and states as described above.

- $\partial^*(p_0, \varepsilon) = \{p_0\}$ so $q_0 = \{p_0\}$.
- $\delta(q_0, a)$ will be $\partial^*(p_0, a)$, which is $\{p_0\}$, so $\delta(q_0, a) = q_0$.
- $\delta(q_0, b)$ will be $\partial^*(p_0, b)$, which is $\{p_0, p_1\}$, so we need to add a new state $q_1 = \{p_0, p_1\}$ to the DFA; and add $\delta(q_0, b) = q_1$ to the DFA's transition function.
- $\delta(q_1, a)$ will be $\partial^*(p_0, a)$ unioned with $\partial^*(p_1, a)$ since $q_1 = \{p_0, p_1\}$. Since $\partial^*(p_0, a) \cup \partial^*(p_1, a) = \{p_0\} \cup \{p_2\} = \{p_0, p_2\}$, we need to add a new state $q_2 = \{p_0, p_2\}$ to the DFA, and a transition $\delta(q_1, a) = q_2$.
- $\delta(q_1, b)$ will be $\partial^*(p_0, b)$ unioned with $\partial^*(p_1, b)$, which gives $\{p_0, p_1\} \cup \{p_2\}$, which again gives us a new state q_3 to add to the DFA, together with the transition $\delta(q_1, b) = q_3$.

At this point, our partially-constructed DFA looks as shown below:

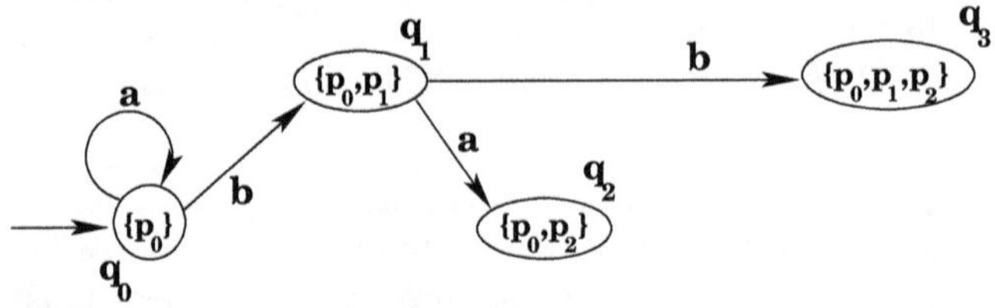

The construction continues as long as there are new states being added, and new transitions from those states that have to be computed. The final DFA is shown below.

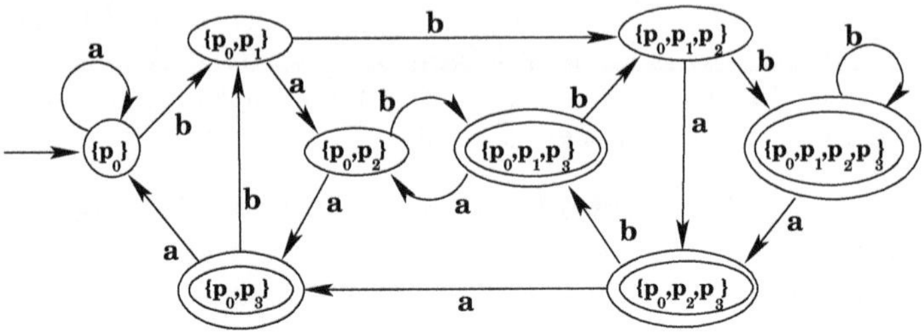

Exercises

1. What language does the NFA in Example 3.17 accept?

2. Give a DFA that accepts the language accepted by the following NFA.

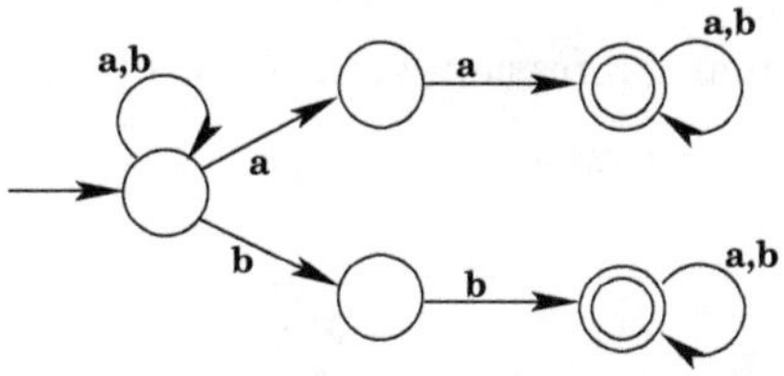

3. Give a DFA that accepts the language accepted by the following NFA. (Be sure to note that, for example, it is possible to reach both q_1 and q_3 from q_0 on consumption of an a, because of the ε-transition.)

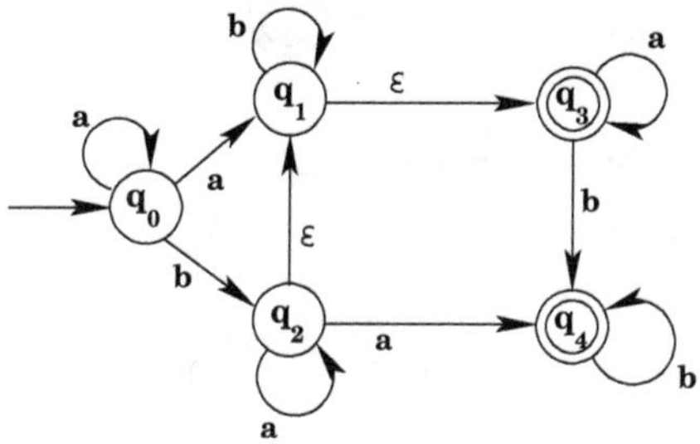

3.6 Finite-State Automata and Regular Languages

We know now that our two models for mechanical language recognition actually recognize the same class of languages. The question still remains: do they recognize the same class of languages as the class generated mechanically by regular expressions? The answer turns out to be "yes". There are two parts to proving this: first that every language generated can be recognized, and second that every language recognized can be generated.

Theorem 3.3. *Every language generated by a regular expression can be recognized by an NFA.*

Proof. The proof of this theorem is a nice example of a proof by induction on the structure of regular expressions. The definition of regular expression is inductive: Φ, ε, and a are the simplest regular expressions, and then more complicated regular expressions can be built from these. We will show that there are NFAs that accept the languages generated by the simplest regular expressions, and then show how those machines can be put together to form machines that accept languages generated by more complicated regular expressions.

Consider the regular expression Φ. $L(\Phi) = \{\}$. Here is a machine that accepts $\{\}$:

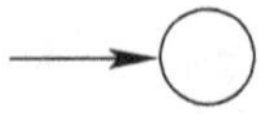

Consider the regular expression ε. $L(\varepsilon) = \{\varepsilon\}$. Here is a machine that accepts $\{\varepsilon\}$:

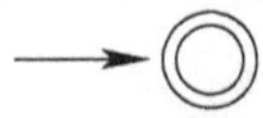

Consider the regular expression a. $L(a) = \{a\}$. Here is a machine that accepts $\{a\}$:

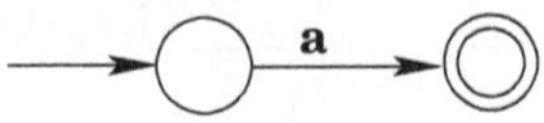

Now suppose that you have NFAs that accept the languages generated by the regular expressions r_1 and r_2. Building a machine that accepts $L(r_1 \mid r_2)$ is fairly straightforward: take an NFA M_1 that accepts $L(r_1)$ and an NFA M_2 that accepts $L(r_2)$. Introduce a new state q_{new}, connect it to the start states of M_1 and M_2 via ε-transitions, and designate it as the start state of the new machine. No other transitions are added. The final states of M_1 together with the final states of M_2 are designated as the final states of the new machine. It should be fairly clear that this new machine accepts exactly those strings accepted by M_1 together with those strings accepted by M_2: any string w that was accepted by M_1 will be accepted by the new NFA by starting with an ε-transition to the old start state of M_1 and then following the accepting path through M_1; similarly, any string accepted by M_2 will be accepted by the new machine; these are the only strings that will be accepted by the new machine, as on any input w all the new machine can do is make an ε-move to M_1's (or M_2's) start state, and from there w will only be accepted by the new machine if it is accepted by M_1 (or M_2). Thus, the new machine accepts $L(M_1) \cup L(M_2)$, which is $L(r_1) \cup L(r_2)$, which is exactly the definition of $L(r_1 \mid r_2)$.

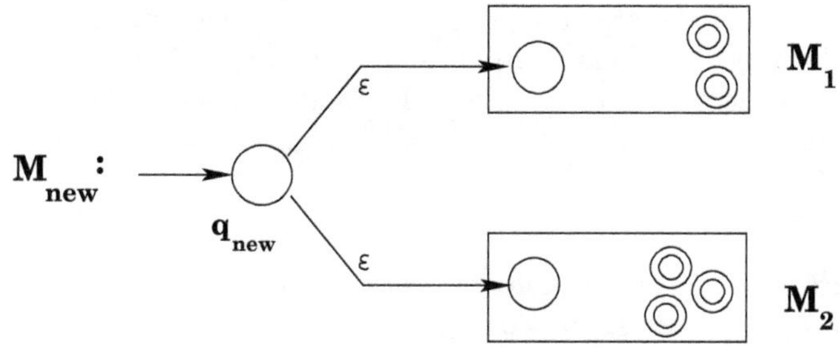

(A pause before we continue: note that for the simplest regular expressions, the machines that we created to accept the languages generated by the regular expressions were in fact DFAs. In our last case above, however, we needed ε-transitions to build the new machine, and so if we were trying to prove that every regular language could be accepted by a DFA, our proof would be in trouble. THIS DOES NOT MEAN that the statement "every regular language can be accepted by a DFA" is false, just that we can't prove it using this kind of argument, and would have to find an alternative proof.)

Suppose you have machines M_1 and M_2 that accept $L(r_1)$ and $L(r_2)$ respectively. To build a machine that accepts $L(r_1)L(r_2)$ proceed as follows. Make the start state q_{01} of M_1 be the start state of the new machine. Make the final states of M_2 be the final states of the new machine. Add ε-transitions from the final states of M_1 to the start state q_{02} of M_2.

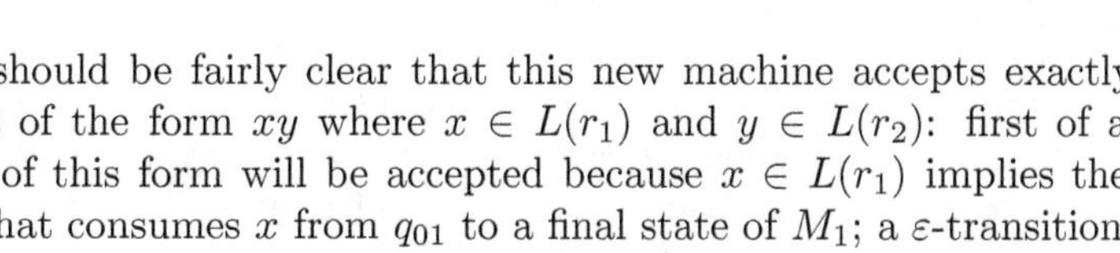

It should be fairly clear that this new machine accepts exactly those strings of the form xy where $x \in L(r_1)$ and $y \in L(r_2)$: first of all, any string of this form will be accepted because $x \in L(r_1)$ implies there is a path that consumes x from q_{01} to a final state of M_1; a ε-transition moves to q_{02}; then $y \in L(r_2)$ implies there is a path that consumes y from q_{02} to a final state of M_2; and the final states of M_2 are the final states of the new machine, so xy will be accepted. Conversely, suppose z is accepted by the new machine. Since the only final states of the new machine are in the old M_2, and the only way to get into M_2 is to take a ε-transition from a final state of M_1, this means that $z = xy$ where x takes the machine from its start state to a final state of M_1, a ε-transition occurs, and then y takes the machine from q_{02} to a final state of M_2. Clearly, $x \in L(r_1)$ and $y \in L(r_2)$.

We leave the construction of an NFA that accepts $L(r^*)$ from an NFA that accepts $L(r)$ as an exercise.

□

Theorem 3.4. *Every language that is accepted by a DFA or an NFA is generated by a regular expression.*

Proving this result is actually fairly involved and not very illuminating. Before presenting a proof, we will give an illustrative example of how one might actually go about extracting a regular expression from an NFA or a DFA. You can go on to read the proof if you are interested.

Example 3.18. Consider the DFA shown below:

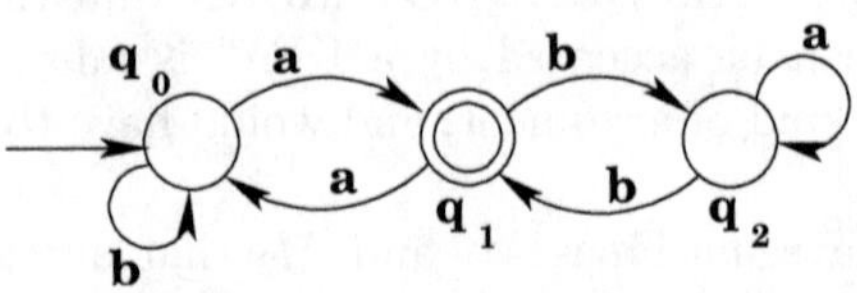

Note that there is a loop from state q_2 back to state q_2: any number of a's will keep the machine in state q_2, and so we label the transition with the regular expression a^*. We do the same thing to the transition labeled b from q_0. (Note that the result is no longer a DFA, but that doesn't concern us, we're just interested in developing a regular expression.)

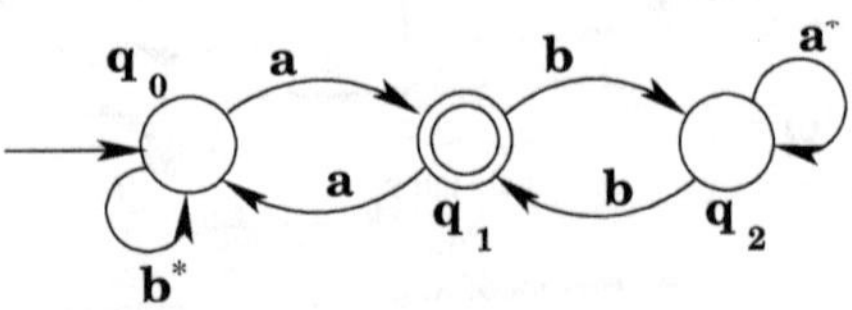

Next we note that there is in fact a loop from q_1 to q_1 via q_0. A regular expression that matches the strings that would move around the loop is ab^*a. So we add a transition labeled ab^*a from q_1 to q_1, and remove the now-irrelevant a-transition from q_1 to q_0. (It is irrelevant because it is not part of any other loop from q_1 to q_1.)

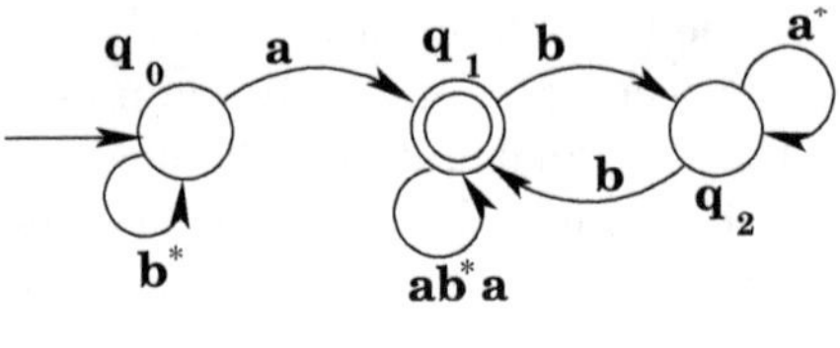

Next we note that there is also a loop from q_1 to q_1 via q_2. A regular expression that matches the strings that would move around the loop is ba^*b. Since the transitions in the loop are the only transitions to or from q_2, we simply remove q_2 and replace it with a transition from q_1 to q_1.

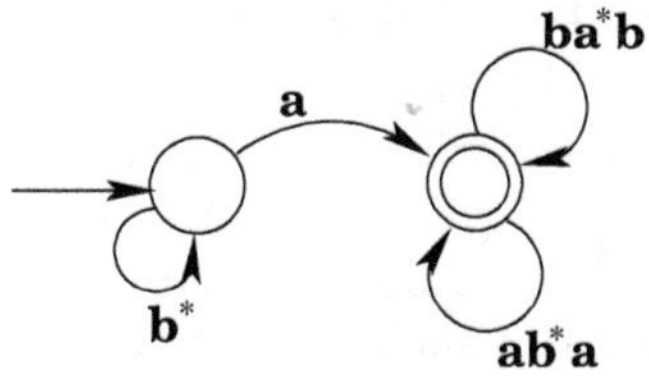

It is now clear from the diagram that strings of the form b^*a get you to state q_1, and any number of repetitions of strings that match ab^*a or ba^*b will keep you there. So the machine accepts $L(b^*a(ab^*a \mid ba^*b)^*)$.

Proof of Theorem 3.4. We prove that the language accepted by a DFA is regular. The proof for NFAs follows from the equivalence between DFAs and NFAs.

Suppose that M is a DFA, where $M = (Q, \Sigma, q_0, \delta, F)$. Let n be the number of states in M, and write $Q = \{q_0, q_1, \ldots, q_{n-1}\}$. We want to consider computations in which M starts in some state q_i, reads a string w, and ends in state q_k. In such a computation, M might go through a series of intermediates states between q_i and q_k:

$$q_i \longrightarrow p_1 \longrightarrow p_2 \cdots \longrightarrow p_r \longrightarrow q_k$$

We are interested in computations in which all of the intermediate states—$p_1, p_2, \ldots, p_r$—are in the set $\{q_0, q_1, \ldots, q_{j-1}\}$, for some number j. We define $R_{i,j,k}$ to be the set of all strings w in Σ^* that are consumed by such a computation. That is, $w \in R_{i,j,k}$ if and only if when M starts in state q_i and reads w, it ends in state q_k, and all the intermediate states between q_i and q_k are in the set $\{q_0, q_1, \ldots, q_{j-1}\}$. $R_{i,j,k}$ is a language over Σ. We show that $R_{i,j,k}$ for $0 \le i < n$, $0 \le j \le n$, $0 \le k < n$.

Consider the language $R_{i,0,k}$. For $w \in R_{i,0,k}$, the set of allowable intermediate states is empty. Since there can be no intermediate states, it follows that there can be at most one step in the computation that starts in state q_i, reads w, and ends in state q_k. So, $|w|$ can be at most one. This means that $R_{i,0,k}$ is finite, and hence is regular. (In fact, $R_{i,0,k} = \{a \in \Sigma \mid \delta(q_i, a) = q_k\}$, for $i \ne k$, and $R_{i,0,i} = \{\varepsilon\} \cup \{a \in \Sigma \mid \delta(q_i, a) = q_i\}$. Note that in many cases, $R_{i,0,k}$ will be the empty set.)

We now proceed by induction on j to show that $R_{i,j,k}$ is regular for all i and k. We have proved the base case, $j = 0$. Suppose that $0 \le j < n$

we already know that $R_{i,j,k}$ is regular for all i and all k. We need to show that $R_{i,j+1,k}$ is regular for all i and k. In fact,

$$R_{i,j+1,k} = R_{i,j,k} \cup \left(R_{i,j,j} R^*_{j,j,j} R_{j,j,k}\right)$$

which is regular because $R_{i,j,k}$ is regular for all i and k, and because the union, concatenation, and Kleene star of regular languages are regular.

To see that the above equation holds, consider a string $w \in \Sigma^*$. Now, $w \in R_{i,j+1,k}$ if and only if when M starts in state q_i and reads w, it ends in state q_k, with all intermediate states in the computation in the set $\{q_0, q_1, \ldots, q_j\}$. Consider such a computation. There are two cases: Either q_j occurs as an intermediate state in the computation, or it does not. If it does **not** occur, then all the intermediate states are in the set $\{q_0, q_1, \ldots, q_{j-1}\}$, which means that in fact $w \in R_{i,j,k}$. If q_j **does** occur as an intermediate state in the computation, then we can break the computation into phases, by dividing it at each point where q_j occurs as an intermediate state. This breaks w into a concatenation $w = xy_1y_2 \cdots y_r z$. The string x is consumed in the first phase of the computation, during which M goes from state q_i to the first occurrence of q_j; since the intermediate states in this computation are in the set $\{q_0, q_1, \ldots, q_{j-1}\}$, $x \in R_{i,j,j}$. The string z is consumed by the last phase of the computation, in which M goes from the final occurrence of q_j to q_k, so that $z \in R_{j,j,k}$. And each string y_t is consumed in a phase of the computation in which M goes from one occurrence of q_j to the next occurrence of q_j, so that $y_r \in R_{j,j,j}$. This means that $w = xy_1y_2 \cdots y_r z \in R_{i,j,j} R^*_{j,j,j} R_{j,j,k}$.

We now know, in particular, that $R_{0,n,k}$ is a regular language for all k. But $R_{0,n,k}$ consists of all strings $w \in \Sigma^*$ such that when M starts in state q_0 and reads w, it ends in state q_k (with **no** restriction on the intermediate states in the computation, since every state of M is in the set $\{q_0, q_1, \ldots, q_{n-1}\}$). To finish the proof that $L(M)$ is regular, it is only necessary to note that

$$L(M) = \bigcup_{q_k \in F} R_{0,n,k}$$

which is regular since it is a union of regular languages. This equation is true since a string w is in $L(M)$ if and only if when M starts in state q_0 and reads w, in ends in some accepting state $q_k \in F$. This is the same as saying $w \in R_{0,n,k}$ for some k with $q_k \in F$. □

We have already seen that if two languages L_1 and L_2 are regular, then so are $L_1 \cup L_2$, L_1L_2, and L_1^* (and of course L_2^*). We have not yet seen, however, how the common set operations intersection and complementation affect regularity. Is the complement of a regular language regular? How about the intersection of two regular languages?

Both of these questions can be answered by thinking of regular languages in terms of their acceptance by DFAs. Let's consider first the question of complementation. Suppose we have an arbitrary regular language L. We know there is a DFA M that accepts L. Pause a moment and try to think of a modification that you could make to M that would produce a new machine M' that accepts $\overline{L}$.... Okay, the obvious thing to try is to make M' be a copy of M with all final states of M becoming non-final states of M' and vice versa. This is in fact exactly right: M' does in fact accept $\overline{L}$. To verify this, consider an arbitrary string w. The transition functions for the two machines M and M' are identical, so $\delta^*(q_0, w)$ is the same state in both M and M'; if that state is accepting in M then it is non-accepting in M', so if w is accepted by M it is not accepted by M'; if the state is non-accepting in M then it is accepting in M', so if w is not accepted by M then it is accepted by M'. Thus M' accepts exactly those strings that M does not, and hence accepts $\overline{L}$.

It is worth pausing for a moment and looking at the above argument a bit longer. Would the argument have worked if we had looked at an arbitrary language L and an arbitrary NFA M that accepted L? That is, if we had built a new machine M' in which the final and non-final states had been switched, would the new NFA M' accept the complement of the language accepted by M? The answer is "not necessarily". Remember that acceptance in an NFA is determined based on whether or not at least one of the states reached by a string is accepting. So any string w with the property that $\partial^*(q_0, w)$ contains both accepting and non-accepting states of M would be accepted both by M and by M'.

Now let's turn to the question of intersection. Given two regular languages L_1 and L_2, is $L_1 \cap L_2$ regular? Again, it is useful to think in terms of DFAs: given machines M_1 and M_2 that accept L_1 and L_2, can you use them to build a new machine that accepts $L_1 \cap L_2$? The answer is yes, and the idea behind the construction bears some resemblance to that behind the NFA-to-DFA construction. We want a new machine where transitions reflect the transitions of both M_1 and M_2 simultaneously, and we want to accept a string w only if that those sequences of transitions lead to final states in both M_1 and M_2. So we associate the states of our new machine M with pairs of states from M_1 and M_2. For each state (q_1, q_2) in the new machine and input symbol a, define $\delta((q_1, q_2), a)$ to be the state $(\delta_1(q_1, a), \delta_2(q_2, a))$. The start state q_0 of M is (q_{01}, q_{02}), where q_{0i} is the start state of M_i. The final states of M are the the states of the form (q_{f1}, q_{f2}) where q_{f1} is an accepting state of M_1 and q_{f2} is an accepting state of M_2. You should convince yourself that M accepts a string x iff x is accepted by both M_1 and M_2.

The results of the previous section and the preceding discussion are summarized by the following theorem:

Theorem 3.5. *The intersection of two regular languages is a regular language.*
The union of two regular languages is a regular language.
The concatenation of two regular languages is a regular language.
The complement of a regular language is a regular language.
The Kleene closure of a regular language is a regular language.

Exercises

1. Give a DFA that accepts the intersection of the languages accepted by the machines shown below. (Suggestion: use the construction discussed in the chapter just before Theorem 3.5.)

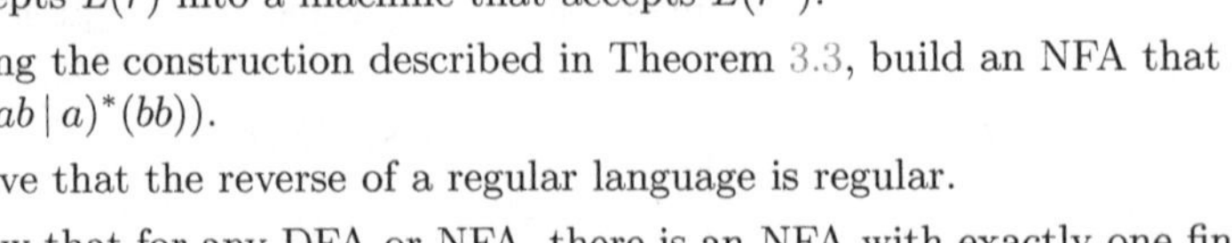

2. Complete the proof of Theorem 3.3 by showing how to modify a machine that accepts $L(r)$ into a machine that accepts $L(r^*)$.
3. Using the construction described in Theorem 3.3, build an NFA that accepts $L((ab\,|\,a)^*(bb))$.
4. Prove that the reverse of a regular language is regular.
5. Show that for any DFA or NFA, there is an NFA with exactly one final state that accepts the same language.
6. Suppose we change the model of NFAs to allow NFAs to have multiple start states. Show that for any "NFA" with multiple start states, there is an NFA with exactly one start state that accepts the same language.
7. Suppose that $M_1 = (Q_1, \Sigma, q_1, \delta_1, F_1)$ and $M_2 = (Q_2, \Sigma, q_2, \delta_2, F_2)$ are DFAs over the alphabet Σ. It is possible to construct a DFA that accepts the langauge $L(M_1) \cap L(M_2)$ in a single step. Define the DFA

$$M = (Q_1 \times Q_2, \Sigma, (q_1, q_2), \delta, F_1 \times F_2)$$

where δ is the function from $(Q_1 \times Q_2) \times \Sigma$ to $Q_1 \times Q_2$ that is defined by: $\delta((p1, p2), \sigma)) = (\delta_1(p_1, \sigma), \delta_2(p_2, \sigma))$. Convince yourself that this definition makes sense. (For example, note that states in M are pairs (p_1, p_2) of states, where $p_1 \in Q_1$ and $p_2 \in Q_2$, and note that the start state (q_1, q_2) in M is in fact a state in M.) Prove that $L(M) = L(M_1) \cap L(M_2)$, and explain why this shows that the intersection of any two regular languages is regular. This proof—if you can get past the notation—is more direct than the one outlined above.

3.7 Non-regular Languages

The fact that our models for mechanical language-recognition accept exactly the same languages as those generated by our mechanical language-generation system would seem to be a very positive indication that in "regular" we have in fact managed to isolate whatever characteristic it is that makes a language "mechanical". Unfortunately, there are languages that we intuitively think of as being mechanically-recognizable (and which we could write C++ programs to recognize) that are not in fact regular.

How does one prove that a language is not regular? We could try proving that there is no DFA or NFA that accepts it, or no regular expression that generates it, but this kind of argument is generally rather difficult to make. It is hard to rule out all possible automata and all possible regular expressions. Instead, we will look at a property that all regular languages have; proving that a given language does not have this property then becomes a way of proving that that language is not regular.

Consider the language $L = \{w \in \{a,b\}^* \mid n_a(w) = 2 \bmod 3,\ n_b(w) = 2 \bmod 3\}$. Below is a DFA that accepts this language, with states numbered 1 through 9.

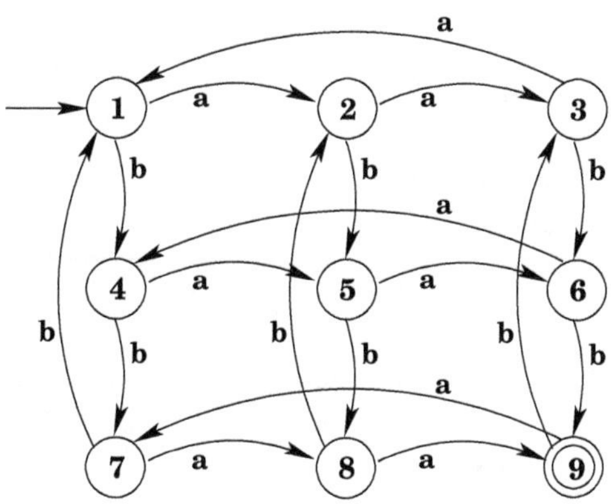

Consider the sequence of states that the machine passes through while processing the string *abbbabb*. Note that there is a repeated state (state 2). We say that *abbbabb* "goes through the state 2 twice", meaning that in the course of the string being processed, the machine is in state 2 twice (at least). Call the section of the string that takes you around the loop y, the preceding section x, and the rest z. Then xz is accepted, $xyyz$ is accepted, $xyyyz$ is accepted, etc. Note that the string *aabb* cannot be divided this way, because it does not go through the same state twice. Which strings **can** be divided this way? Any string that goes through the same state twice. This may include some relatively short strings and must include any string with length greater than or equal to 9, because there are only 9 states in the machine, and so repetition must occur after 9 input symbols at the latest.

More generally, consider an arbitrary DFA M, and let the number of states in M be n. Then any string w that is accepted by M and has n or more symbols must go through the same state twice, and can therefore be broken up into three pieces x, y, z (where y contains at least one symbol) so that $w = xyz$ and

xz is accepted by M
xyz is accepted by M (after all, we started with w in $L(M)$)
$xyyz$ is accepted by M
etc.

Note that you can actually say even more: within the first n characters of w you must already get a repeated state, so you can always find an x, y, z as described above where, in addition, the xy portion of w (the portion of w that takes you to and back to a repeated state) contains at most n symbols.

So altogether, if M is an n-state DFA that accepts L, and w is a string in L whose length is at least n, then w can be broken down into three pieces x, y, and z, $w = xyz$, such that

(i) x and y together contain no more than n symbols;
(ii) y contains at least one symbol;
(iii) xz is accepted by M
  (xyz is accepted by M)
  $xyyz$ is accepted by M
  etc.

The usually-stated form of this result is the Pumping Lemma:

Theorem 3.6. *If L is a regular language, then there is some number $n > 0$ such that any string w in L whose length is greater than or equal to n can be broken down into three pieces x, y, and z, $w = xyz$, such that*

(i) x and y together contain no more than n symbols;

(ii) y contains at least one symbol;

(iii) xz is accepted by M

(xyz is accepted by M)

xyyz is accepted by M

etc.

Though the Pumping Lemma says something about regular languages, it is not used to prove that languages are regular. It says "**if** a language is regular, then certain things happen", not "if certain things happen, **then** you can conclude that the language is regular." However, the Pumping Lemma is useful for proving that languages are not regular, since the contrapositive of "if a language is regular then certain things happen" is "if certain things don't happen then you can conclude that the language is not regular." So what are the "certain things"? Basically, the P.L. says that if a language is regular, there is some "threshold" length for strings, and every string that goes over that threshold can be broken down in a certain way. Therefore, if we can show that "there is some threshold length for strings such that every string that goes over that threshold can be broken down in a certain way" is a false assertion about a language, we can conclude that the language is not regular. How do you show that there is no threshold length? Saying a number is a threshold length for a language means that every string in the language that is at least that long can be broken down in the ways described. So to show that a number is not a threshold value, we have to show that there is some string in the language that is at least that long that cannot be broken down in the appropriate way.

Theorem 3.7. $\{a^n b^n \mid n \geq 0\}$ *is not regular.*

Proof. We do this by showing that there is no threshold value for the language. Let N be an arbitrary candidate for threshold value. We want to show that it is not in fact a threshold value, so we want to find a string in the language whose length is at least N and which can't be broken down in the way described by the Pumping Lemma. What string should we try to prove unbreakable? We can't pick strings like $a^{100}b^{100}$ because we're working with an arbitrary N i.e. making no assumptions about N's value; picking $a^{100}b^{100}$ is implicitly assuming that N is no bigger than 200 — for larger values of N, $a^{100}b^{100}$ would not be "a string whose length is at least N". Whatever string we pick, we **have** to be sure that its length is at least N, no matter what number N is. So we pick, for instance, $w = a^N b^N$. This string is in the language, and its length is at least N, no matter what number N is. If we can show that this string can't be broken down as described by the Pumping Lemma, then we'll have shown that N doesn't work as a threshold value, and since N was an arbitrary number, we will have shown that there is no threshold value for L and hence L is not regular. So let's show that $w = a^N b^N$ can't be broken down appropriately.

We need to show that you can't write $w = a^N b^N$ as $w = xyz$ where x and y together contain at most N symbols, y isn't empty, and all the strings xz, $xyyz$, $xyyyz$, etc. are still in L, i.e. of the form $a^n b^n$ for some number n. The best way to do this is to show that any choice for y (with x being whatever precedes it and z being whatever follows) that satisfies the first two requirements fails to satisfy the third. So what are our possible choices for y? Well, since x and y together can contain at most N symbols, and w starts with N a's, both x and y must be made up entirely of a's; since y can't be empty, it must contain at least one a and (from (i)) no more than N a's. So the possible choices for y are $y = a^k$ for some $1 \leq k \leq N$. We want to show now that none of these choices will satisfy the third requirement by showing that for any value of k, at least one of the strings xz, $xyyz$, $xyyyz$, etc will not be in L. No matter what value we try for k, we don't have to look far for our rogue string: the string xz, which is $a^N b^N$ with k a's deleted from it, looks like $a^{N-k} b^N$, which is clearly not of the form $a^n b^n$. So the only y's that satisfy (i) and (ii) don't satisfy (iii); so w can't be broken down as required; so N is not a threshold value for L; and since N was an arbitrary number, there is no threshold value for L; so L is not regular. □

The fact that languages like $\{a^n b^n \mid n \geq 0\}$ and $\{a^p \mid p \text{ is prime}\}$ are not regular is a severe blow to any idea that regular expressions or finite-state automata capture the language-generation or language-recognition capabilities of a computer: They are both languages that we could easily write programs to recognize. It is not clear how the expressive power of regular expressions could be increased, nor how one might modify the FSA model to obtain a more powerful one. However, in the next chapter you will be introduced to the concept of a *grammar* as a tool for generating languages. The simplest grammars still only produce regular languages, but you will see that more complicated grammars have the power to generate languages far beyond the realm of the regular.

Exercises

1. Use the Pumping Lemma to show that the following languages over $\{a, b\}$ are not regular.

a) $L_1 = \{x \mid n_a(x) = n_b(x)\}$
b) $L_2 = \{xx \mid x \in \{a, b\}^*\}$
c) $L_3 = \{xx^R \mid x \in \{a, b\}^*\}$
d) $L_4 = \{a^n b^m \mid n < m\}$

Chapter 4

Grammars

Both natural languages, such as English and the artificial languages used for programming have a structure known as grammar or syntax. In order to form legal sentences or programs, the parts of the language must be fit together according to certain rules. For natural languages, the rules are somewhat informal (although high-school English teachers might have us believe differently). For programming languages, the rules are absolute, and programs that violate the rules will be rejected by a compiler.

In this chapter, we will study formal grammars and languages defined by them. The languages we look at will, for the most part, be "toy" languages, compared to natural languages or even to programming languages, but the ideas and techniques are basic to any study of language. In fact, many of the ideas arose almost simultaneously in the 1950s in the work of linguists who were studying natural language and programmers who were looking for ways to specify the syntax of programming languages.

The grammars in this chapter are ***generative grammars***. A generative grammar is a set of rules that can be used to generate all the legal strings in a language. We will also consider the closely related idea of ***parsing***. To parse a string means to determine how that string can be generated according to the rules.

This chapter is a continuation of the preceding chapter. Like a regular expression, a grammar is a way to specify a possibly infinite language with a finite amount of information. In fact, we will see that every regular language can be specified by a certain simple type of grammar. We will also see that some languages that can be specified by grammars are not regular.

4.1 Context-free Grammars

In its most general form, a grammar is a set of ***rewriting rules***. A rewriting rule specifies that a certain string of symbols can be substituted for all or part of another string. If w and u are strings, then $w \longrightarrow u$ is a rewriting rule that specifies that the string w can be replaced by the string u. The symbol "$\longrightarrow$" is read "can be rewritten as." Rewriting rules are also called ***production rules*** or ***productions***, and "$\longrightarrow$" can also be read as "produces." For example, if we consider strings over the alphabet $\{a, b, c\}$, then the production rule $aba \longrightarrow cc$ can be applied to the string $abbabac$ to give the string $abbccc$. The substring aba in the string $abbabac$ has been replaced with cc.

In a ***context-free grammar***, every rewriting rule has the form $A \longrightarrow w$, where A is single symbol and w is a string of zero or more symbols. (The grammar is "context-free" in the sense that w can be substituted for A wherever A occurs in a string, regardless of the surrounding context in which A occurs.) The symbols that occur on the left-hand sides of production rules in a context-free grammar are called ***non-terminal symbols***. By convention, the non-terminal symbols are usually uppercase letters. The strings on the right-hand sides of the production rules can include non-terminal symbols as well as other symbols, which are called ***terminal symbols***. By convention, the terminal symbols are usually lowercase letters. Here are some typical production rules that might occur in context-free grammars:

$$A \longrightarrow aAbB$$
$$S \longrightarrow SS$$
$$C \longrightarrow Acc$$
$$B \longrightarrow b$$
$$A \longrightarrow \varepsilon$$

In the last rule in this list, ε represents the empty string, as usual. For example, this rule could be applied to the string $aBaAcA$ to produce the string $aBacA$. The first occurrence of the symbol A in $aBaAcA$ has been replaced by the empty string—which is just another way of saying that the symbol has been dropped from the string.

In every context-free grammar, one of the non-terminal symbols is designated as the ***start symbol*** of the grammar. The start symbol is often, though not always, denoted by S. When the grammar is used to generate strings in a language, the idea is to start with a string consisting of nothing but the start symbol. Then a sequence of production rules is applied. Each application of a production rule to the string transforms the string to a new string. If and when this process produces a string that consists purely of terminal symbols, the process ends. That string of terminal symbols is one of the strings in the language generated by the grammar. In fact, the language consists precisely of all strings of terminal symbols that can be produced in this way.

As a simple example, consider a grammar that has three production rules: $S \longrightarrow aS$, $S \longrightarrow bS$, and $S \longrightarrow b$. In this example, S is the only non-terminal symbol, and the terminal symbols are a and b. Starting from the string S, we can apply any of the three rules of the grammar to produce either aS, bS, or b. Since the string b contains no non-terminals, we see that b is one of the strings in the language generated by this grammar. The strings aS and bS are not in that language, since they contain the non-terminal symbol S, but we can continue to apply production rules to these strings. From aS, for example, we can obtain aaS, abS, or ab. From abS, we go on to obtain $abaS$, $abbS$, or abb. The strings ab and abb are in the language generated by the grammar. It's not hard to see that any string of a's and b's that ends with a b can be generated by this grammar, and that these are the only strings that can be generated. That is, the language generated by this grammar is the regular language specified by the regular expression $(a \mid b)^*b$.

It's time to give some formal definitions of the concepts which we have been discussing.

Definition 4.1. A ***context-free grammar*** is a 4-tuple (V, Σ, P, S), where:

1. V is a finite set of symbols. The elements of V are the non-terminal symbols of the grammar.
2. Σ is a finite set of symbols such that $V \cap \Sigma = \emptyset$. The elements of Σ are the terminal symbols of the grammar.
3. P is a set of production rules. Each rule is of the form $A \longrightarrow w$ where A is one of the symbols in V and w is a string in the language $(V \cup \Sigma)^*$.
4. $S \in V$. S is the start symbol of the grammar.

Even though this is the formal definition, grammars are often specified informally simply by listing the set of production rules. When this is done it is assumed, unless otherwise specified, that the non-terminal symbols are just the symbols that occur on the left-hand sides of production rules of the grammar. The terminal symbols are all the other symbols that occur on the right-hand sides of production rules. The start symbol is the symbol that occurs on the left-hand side of the first production rule in the list.

Thus, the list of production rules

$$\begin{aligned}
T &\longrightarrow TT \\
T &\longrightarrow A \\
A &\longrightarrow aAa \\
A &\longrightarrow bB \\
B &\longrightarrow bB \\
B &\longrightarrow \varepsilon
\end{aligned}$$

specifies a grammar $G = (V, \Sigma, P, T)$ where V is $\{T, A, B\}$, Σ is $\{a, b\}$, and T is the start symbol. P, of course, is a set containing the six production rules in the list.

Let $G = (V, \Sigma, P, S)$ be a context-free grammar. Suppose that x and y are strings in the language $(V \cup \Sigma)^*$. The notation $x \Longrightarrow_G y$ is used to express the fact that y can be obtained from x by applying one of the production rules in P. To be more exact, we say that $x \Longrightarrow_G y$ if and only if there is a production rule $A \longrightarrow w$ in the grammar and two strings u and v in the language $(V \cup \Sigma)^*$ such that $x = uAv$ and $y = uwv$. The fact that $x = uAv$ is just a way of saying that A occurs somewhere in x. When the production rule $A \longrightarrow w$ is applied to substitute w for A in uAv, the result is uwv, which is y. Note that either u or v or both can be the empty string.

If a string y can be obtained from a string x by applying a sequence of zero or more production rules, we write $x \Longrightarrow_G^* y$. In most cases, the "G" in the notations $\Longrightarrow_G$ and $\Longrightarrow_G^*$ will be omitted, assuming that the grammar in question is understood. Note that $\Longrightarrow$ is a relation on the set $(V \cup \Sigma)^*$. The relation $\Longrightarrow^*$ is the reflexive, transitive closure of that relation. (This explains the use of "$*$", which is usually used to denote the transitive, but not necessarily reflexive, closure of a relation. In this case, $\Longrightarrow^*$ is reflexive as well as transitive since $x \Longrightarrow^* x$ is true for any string x.) For example, using the grammar that is defined by the above list of production rules, we have

$$\begin{aligned}
aTB &\Longrightarrow aTTB \\
&\Longrightarrow aTAB \\
&\Longrightarrow aTAbB \\
&\Longrightarrow aTbBbB \\
&\Longrightarrow aTbbB
\end{aligned}$$

From this, it follows that $aTB \Longrightarrow^* aTbbB$. The relation $\Longrightarrow$ is read "yields" or "produces" while $\Longrightarrow^*$ can be read "yields in zero or more steps" or "produces in zero or more steps." The following theorem states some simple facts about the relations $\Longrightarrow$ and $\Longrightarrow^*$:

Theorem 4.1. *Let G be the context-free grammar (V, Σ, P, S). Then:*

1. *If x and y are strings in $(V \cup \Sigma)^*$ such that $x \Longrightarrow y$, then $x \Longrightarrow^* y$.*
2. *If x, y, and z are strings in $(V \cup \Sigma)^*$ such that $x \Longrightarrow^* y$ and $y \Longrightarrow^* z$, then $x \Longrightarrow^* z$.*
3. *If x and y are strings in $(V \cup \Sigma)^*$ such that $x \Longrightarrow y$, and if s and t are any strings in $(V \cup \Sigma)^*$, then $sxt \Longrightarrow syt$.*
4. *If x and y are strings in $(V \cup \Sigma)^*$ such that $x \Longrightarrow^* y$, and if s and t are any strings in $(V \cup \Sigma)^*$, then $sxt \Longrightarrow^* syt$.*

Proof. Parts 1 and 2 follow from the fact that $\Longrightarrow^*$ is the transitive closure of $\Longrightarrow$. Part 4 follows easily from Part 3. (I leave this as an exercise.) To prove Part 3, suppose that x, y, s, and t are strings such that $x \Longrightarrow y$. By definition, this means that there exist strings u and v and a production rule $A \longrightarrow w$ such that $x = uAv$ and $y = uwv$. But then we also have $sxt = suAvt$ and $syt = suwvt$. These two equations, along with the existence of the production rule $A \longrightarrow w$ show, by definition, that $sxt \Longrightarrow syt$. □

We can use $\Longrightarrow^*$ to give a formal definition of the language generated by a context-free grammar:

Definition 4.2. Suppose that $G = (V, \Sigma, P, S)$ is a context-free grammar. Then the language generated by G is the language $L(G)$ over the alphabet Σ defined by

$$L(G) = \{w \in \Sigma^* \mid S \Longrightarrow_G^* w\}$$

That is, $L(G)$ contains any string of terminal symbols that can be obtained by starting with the string consisting of the start symbol, S, and applying a sequence of production rules.

A language L is said to be a ***context-free language*** if there is a context-free grammar G such that $L(G)$ is L. Note that there might be many different context-free grammars that generate the same context-free language. Two context-free grammars that generate the same language are said to be ***equivalent***.

Suppose G is a context-free grammar with start symbol S and suppose $w \in L(G)$. By definition, this means that there is a sequence of one or more applications of production rules which produces the string w from S. This sequence has the form $S \Longrightarrow x_1 \Longrightarrow x_2 \Longrightarrow \cdots \Longrightarrow w$. Such a sequence is called a ***derivation*** of w (in the grammar G). Note that w might have more than one derivation. That is, it might be possible to produce w in several different ways.

Consider the language $L = \{a^n b^n \mid n \in \mathbb{N}\}$. We already know that L is not a regular language. However, it is a context-free language. That is, there is a context-free grammar such that L is the language generated by G. This gives us our first theorem about grammars:

Theorem 4.2. *Let L be the language $L = \{a^n b^n \mid n \in \mathbb{N}\}$. Let G be the context-free grammar (V, Σ, P, S) where $V = \{S\}$, $\Sigma = \{a, b\}$ and P consists of the productions*

$$S \longrightarrow aSb$$
$$S \longrightarrow \varepsilon$$

Then $L = L(G)$, so that L is a context-free language. In particular, there exist context-free languages which are not regular.

Proof. To show that $L = L(G)$, we must show both that $L \subseteq L(G)$ and that $L(G) \subseteq L$. To show that $L \subseteq L(G)$, let w be an arbitrary element of L. By definition of L, $w = a^n b^n$ for some $n \in \mathbb{N}$. We show that $w \in L(G)$ by induction on n. In the case where $n = 0$, we have $w = \varepsilon$. Now, $\varepsilon \in L(G)$ since ε can be produced from the start symbol S by an application of the rule $S \longrightarrow \varepsilon$, so our claim is true for $n = 0$. Now, suppose that $k \in \mathbb{N}$ and that we already know that $a^k b^k \in L(G)$. We must show that $a^{k+1} b^{k+1} \in L(G)$. Since $S \Longrightarrow^* a^k b^k$, we also have, by Theorem 4.1, that $aSb \Longrightarrow^* aa^k b^k b$. That is, $aSb \Longrightarrow^* a^{k+1} b^{k+1}$. Combining this with the production rule $S \longrightarrow aSb$, we see that $S \Longrightarrow^* a^{k+1} b^{k+1}$. This means that $a^{k+1} b^{k+1} \in L(G)$, as we wanted to show. This completes the proof that $L \subseteq L(G)$.

To show that $L(G) \subseteq L$, suppose that $w \in L(G)$. That is, $S \Longrightarrow^* w$. We must show that $w = a^n b^n$ for some n. Since $S \Longrightarrow^* w$, there is a derivation $S \Longrightarrow x_0 \Longrightarrow x_1 \Longrightarrow \cdots \Longrightarrow x_n$, where $w = x_n$. We first prove by induction on n that in any derivation $S \Longrightarrow x_0 \Longrightarrow x_1 \Longrightarrow \cdots \Longrightarrow x_n$, we must have either $x_n = a^n b^n$ or $x_n = a^{n+1} S b^{n+1}$. Consider the case $n = 0$. Suppose $S \Longrightarrow x_0$. Then, we must have that $S \longrightarrow x_0$ is a rule in the grammar, so x_0 must be either ε or aSb. Since $\varepsilon = a^0 b^0$ and $aSb = a^{0+1} S b^{0+1}$, x_0 is of the required form. Next, consider the inductive case. Suppose that $k > 1$ and we already know that in any derivation $S \Longrightarrow x_0 \Longrightarrow x_1 \Longrightarrow \cdots \Longrightarrow x_k$, we must have $x_k = a^k b^k$ or $x = a^{k+1} S b^{k+1}$. Suppose that $S \Longrightarrow x_0 \Longrightarrow x_1 \Longrightarrow \cdots \Longrightarrow x_k \Longrightarrow x_{k+1}$. We know by induction that $x_k = a^k b^k$ or $x = a^{k+1} S b^{k+1}$, but since $x_k \Longrightarrow x_{k+1}$ and $a^k b^k$ contains no non-terminal symbols, we must have $x_k = a^{k+1} S b^{k+1}$. Since x_{k+1} is obtained by applying one of the production rules $S \longrightarrow \varepsilon$ or $S \longrightarrow aSb$ to x_k, x_{k+1} is either $a^{k+1} \varepsilon b^{k+1}$ or $a^{k+1} aSbb^{k+1}$. That is, x_{k+1} is either $a^{k+1} b^{k+1}$ or $a^{k+2} S b^{k+2}$, as we wanted to show. This completes the induction. Turning back to w, we see that w must be of the form $a^n b^n$ or of the form $a^n S b^n$. But since $w \in L(G)$, it can contain no non-terminal symbols, so w must be of the form $a^n b^n$, as we wanted to show. This completes the proof that $L(G) \subseteq L$. □

I have given a very formal and detailed proof of this theorem, to show how it can be done and to show how induction plays a role in many proofs about grammars. However, a more informal proof of the theorem would probably be acceptable and might even be more convincing. To show that $L \subseteq L(G)$, we could just note that the derivation $S \Longrightarrow aSb \Longrightarrow a^2Sb^2 \Longrightarrow \cdots \Longrightarrow a^nSb^n \Longrightarrow a^nb^n$ demonstrates that $a^nb^n \in L$. On the other hand, it is clear that every derivation for this grammar must be of this form, so every string in $L(G)$ is of the form a^nb^n.

For another example, consider the language $\{a^nb^m \mid n \geq m \geq 0\}$. Let's try to design a grammar that generates this language. This is similar to the previous example, but now we want to include strings that contain more a's than b's. The production rule $S \longrightarrow aSb$ always produces the same number of a's and b's. Can we modify this idea to produce more a's than b's?

One approach would be to produce a string containing just as many a's as b's, and then to add some extra a's. A rule that can generate any number of a's is $A \longrightarrow aA$. After applying the rule $S \longrightarrow aSb$ for a while, we want to move to a new state in which we apply the rule $A \longrightarrow aA$. We can get to the new state by applying a rule $S \longrightarrow A$ that changes the S into an A. We still need a way to finish the process, which means getting rid of all non-terminal symbols in the string. For this, we can use the rule $A \longrightarrow \varepsilon$. Putting these rules together, we get the grammar

$$\begin{aligned} S &\longrightarrow aSb \\ S &\longrightarrow A \\ A &\longrightarrow aA \\ A &\longrightarrow \varepsilon \end{aligned}$$

This grammar does indeed generate the language $\{a^nb^m \mid n \geq m \geq 0\}$. With slight variations on this grammar, we can produce other related languages. For example, if we replace the rule $A \longrightarrow \varepsilon$ with $A \longrightarrow a$, we get the language $\{a^nb^m \mid n > m \geq 0\}$.

There are other ways to generate the language $\{a^nb^m \mid n \geq m \geq 0\}$. For example, the extra non-terminal symbol, A, is not really necessary, if we allow S to sometimes produce a single a without a b. This leads to the grammar

$$S \longrightarrow aSb$$
$$S \longrightarrow aS$$
$$S \longrightarrow \varepsilon$$

(But note that the rule $S \longrightarrow Sa$ would not work in place of $S \longrightarrow aS$, since it would allow the production of strings in which an a can follow a b, and there are no such strings in the language $\{a^n b^m \mid n \geq m \geq 0\}$.) And here are two more grammars that generate this language:

$S \longrightarrow AB$	$S \longrightarrow ASb$
$A \longrightarrow aA$	$A \longrightarrow aA$
$B \longrightarrow aBb$	$S \longrightarrow \varepsilon$
$A \longrightarrow \varepsilon$	$A \longrightarrow \varepsilon$
$B \longrightarrow \varepsilon$	

Consider another variation on the language $\{a^n b^n \mid n \in \mathbb{N}\}$, in which the a's and b's can occur in any order, but the number of a's is still equal to the number of b's. This language can be defined as $L = \{w \in \{a,b\}^* \mid n_a(w) = n_b(w)\}$. This language includes strings such as *abbaab*, *baab*, and *bbbaaa*.

Let's start with the grammar containing the rules $S \longrightarrow aSb$ and $S \longrightarrow \varepsilon$. We can try adding the rule $S \longrightarrow bSa$. Every string that can be generated using these three rules is in the language L. However, not every string in L can be generated. A derivation that starts with $S \Longrightarrow aSb$ can only produce strings that begin with a and end with b. A derivation that starts with $S \Longrightarrow bSa$ can only generate strings that begin with b and end with a. There is no way to generate the strings *baab* or *abbbabaaba*, which are in the language L. But we shall see that any string in L that begins and ends with the same letter can be written in the form xy where x and y are shorter strings in L. To produce strings of this form, we need one more rule, $S \longrightarrow SS$. The complete set of production rules for the language L is

$$S \longrightarrow aSb$$
$$S \longrightarrow bSa$$
$$S \longrightarrow SS$$
$$S \longrightarrow \varepsilon$$

It's easy to see that every string that can be generated using these rules is in L, since each rule introduces the same number of a's as b's. But we also need to check that every string w in L can be generated by these rules.

This can be done by induction on the length of w, using the second form of the principle of mathematical induction. In the base case, $|w| = 0$ and $w = \varepsilon$. In this case, $w \in L$ since $S \Longrightarrow \varepsilon$ in one step. Suppose $|w| = k$, where $k > 0$, and suppose that we already know that for any $x \in L$ with $|x| < k$, $S \Longrightarrow^* x$. To finish the induction we must show, based on this induction hypothesis, that $S \Longrightarrow^* w$.

Suppose that the first and last characters of w are different. Then w is either of the form axb or of the form bxa, for some string x. Let's assume that w is of the form axb. (The case where w is of the form bxa is handled in a similar way.) Since w has the same number of a's and b's and since x has one fewer a than w and one fewer b than w, x must also have the same number of a's as b's. That is $x \in L$. But $|x| = |w| - 2 < k$, so by the induction hypothesis, $x \in L(G)$. So we have $S \Longrightarrow^* x$. By Theorem 4.1, we get then $aSb \Longrightarrow^* axb$. Combining this with the fact that $S \Longrightarrow aSb$, we get that $S \Longrightarrow^* axb$, that is, $S \Longrightarrow^* w$. This proves that $w \in L(G)$.

Finally, suppose that the first and last characters of w are the same. Let's say that w begins and ends with a. (The case where w begins and ends with b is handled in a similar way.) I claim that w can be written in the form xy where $x \in L(G)$ and $y \in L(G)$ and neither x nor y is the empty string. This will finish the induction, since we will then have by the induction hypothesis that $S \Longrightarrow^* x$ and $S \Longrightarrow^* y$, and we can derive xy from S by first applying the rule $S \longrightarrow SS$ and then using the first S on the right-hand side to derive x and the second to derive y.

It only remains to figure out how to divide w into two strings x and y which are both in $L(G)$. The technique that is used is one that is more generally useful. Suppose that $w = c_1c_2 \cdots c_k$, where each c_i is either a or b. Consider the sequence of integers r_1, r_2, ..., r_k where for each $i = 1, 2, \dots, k$, r_i is the number of a's in $c_1c_2 \cdots c_i$ minus the number of b's in $c_1c_2 \cdots c_i$. Since $c_1 = a$, $r_1 = 1$. Since $w \in L$, $r_k = 0$. And since $c_k = a$, we must have $r_{k-1} = r_k - 1 = -1$. Furthermore the difference between r_{i+1} and r_i is either 1 or -1, for $i = 1, 2, \dots, k-1$.

Since $r_1 = 1$ and $r_{k-1} = -1$ and the value of r_i goes up or down by 1 when i increases by 1, r_i must be zero for some i between 1 and $k - 1$. That is, r_i cannot get from 1 to -1 unless it passes through zero. Let i be a number between 1 and $k - 1$ such that $r_i = 0$. Let $x = c_1c_2 \cdots c_i$ and let $y = c_{i+1}c_{i+2} \cdots c_k$. Note that $xy = w$. The fact that $r_i = 0$ means that the string $c_1c_2 \cdots c_i$ has the same number of a's and b's, so $x \in L(G)$. It follows automatically that $y \in L(G)$ also. Since i is strictly between 1 and $k - 1$, neither x nor y is the empty string. This is all that we needed to show to finish the proof that $L = L(G)$.

The basic idea of this proof is that if w contains the same number of a's as b's, then an a at the beginning of w must have a "matching" b somewhere in w. This b matches the a in the sense that the corresponding r_i is zero, and the b marks the end of a string x which contains the same number of a's as b's. For example, in the string $aababbabba$, the a at the beginning of the string is matched by the third b, since $aababb$ is the shortest prefix of $aababbabba$ that has an equal number of a's and b's.

Closely related to this idea of matching a's and b's is the idea of ***balanced parentheses***. Consider a string made up of parentheses, such as `(()(()))(())`. The parentheses in this sample string are balanced because each left parenthesis has a matching right parenthesis, and the matching pairs are properly nested. A careful definition uses the sort of integer sequence introduced in the above proof. Let w be a string of parentheses. Write $w = c_1c_2 \cdots c_n$, where each c_i is either `(` or `)`. Define a sequence of integers $r_1, r_2, \ldots, r_n$, where r_i is the number of left parentheses in $c_1c_2 \cdots c_i$ minus the number of right parentheses. We say that the parentheses in w are balanced if $r_n = 0$ and $r_i \geq 0$ for all $i = 1, 2, \ldots, n$. The fact that $r_n = 0$ says that w contains the same number of left parentheses as right parentheses. The fact the $r_i \geq 0$ means that the nesting of pairs of parentheses is correct: You can't have a right parenthesis unless it is balanced by a left parenthesis in the preceding part of the string. The language that consists of all balanced strings of parentheses is context-free. It is generated by the grammar

$$S \longrightarrow (\, S\,)$$
$$S \longrightarrow SS$$
$$S \longrightarrow \varepsilon$$

The proof is similar to the preceding proof about strings of a's and b's. (It might seem that I've made an awfully big fuss about matching and balancing. The reason is that this is one of the few things that we can do with context-free languages that we can't do with regular languages.)

Before leaving this section, we should look at a few more general results. Since we know that most operations on regular languages produce languages that are also regular, we can ask whether a similar result holds for context-free languages. We will see later that the intersection of two context-free languages is not necessarily context-free. Also, the complement of a context-free language is not necessarily context-free. However, some other operations on context-free languages do produce context-free languages.

Theorem 4.3. *Suppose that L and M are context-free languages. Then the languages $L \cup M$, LM, and L^* are also context-free.*

Proof. I will prove only the first claim of the theorem, that $L \cup M$ is context-free. In the exercises for this section, you are asked to construct grammars for LM and L^* (without giving formal proofs that your answers are correct).

Let $G = (V, \Sigma, P, S)$ and $H = (W, \Gamma, Q, T)$ be context-free grammars such that $L = L(G)$ and $M = L(H)$. We can assume that $W \cap V = \emptyset$, since otherwise we could simply rename the non-terminal symbols in W. The idea of the proof is that to generate a string in $L \cup M$, we first decide whether we want a string in L or a string in M. Once that decision is made, to make a string in L, we use production rules from G, while to make a string in M, we use rules from H. We have to design a grammar, K, to represent this process.

Let R be a symbol that is not in any of the alphabets V, W, Σ, or Γ. R will be the start symbol of K. The production rules for K consist of all the production rules from G and H together with two new rules:

$$R \longrightarrow S$$
$$R \longrightarrow T$$

Formally, K is defined to be the grammar

$$(V \cup W \cup \{R\}, P \cup Q \cup \{R \longrightarrow S, R \longrightarrow T\}, \Sigma \cup \Gamma, R)$$

Suppose that $w \in L$. That is $w \in L(G)$, so there is a derivation $S \Longrightarrow_G^* w$. Since every rule from G is also a rule in K, if follows that $S \Longrightarrow_K^* w$. Combining this with the fact that $R \Longrightarrow_K S$, we have that $R \Longrightarrow_K^* w$, and $w \in L(K)$. This shows that $L \subseteq L(K)$. In an exactly similar way, we can show that $M \subseteq L(K)$. Thus, $L \cup M \subseteq L(K)$.

It remains to show that $L(K) \subseteq L \cup M$. Suppose $w \in L(K)$. Then there is a derivation $R \Longrightarrow_K^* w$. This derivation must begin with an application of one of the rules $R \longrightarrow S$ or $R \longrightarrow T$, since these are the only rules in which R appears. If the first rule applied in the derivation is $R \longrightarrow S$, then the remainder of the derivation shows that $S \Longrightarrow_K^* w$. Starting from S, the only rules that can be applied are rules from G, so in fact we have $S \Longrightarrow_G^* w$. This shows that $w \in L$. Similarly, if the first rule applied in the derivation $R \Longrightarrow_K^* w$ is $R \longrightarrow T$, then $w \in M$. In any case, $w \in L \cup M$. This proves that $L(K) \subseteq L \cup M$. □

Finally, we should clarify the relationship between context-free languages and regular languages. We have already seen that there are context-free languages which are not regular. On the other hand, it turns out that every regular language is context-free. That is, given any regular language, there is a context-free grammar that generates that language. This means that any syntax that can be expressed by a regular expression, by a DFA, or by an NFA could also be expressed by a context-free grammar. In fact, we only need a certain restricted type of context-free grammar to duplicate the power of regular expressions.

Definition 4.3. A ***right-regular grammar*** is a context-free grammar in which the right-hand side of every production rule has one of the following forms: the empty string; a string consisting of a single non-terminal symbol; or a string consisting of a single terminal symbol followed by a single non-terminal symbol.

Examples of the types of production rule that are allowed in a right-regular grammar are $A \longrightarrow \varepsilon$, $B \longrightarrow C$, and $D \longrightarrow aE$. The idea of the proof is that given a right-regular grammar, we can build a corresponding NFA and *vice-versa.* The states of the NFA correspond to the non-terminal symbols of the grammar. The start symbol of the grammar corresponds to the starting state of the NFA. A production rule of the form $A \longrightarrow bC$ corresponds to a transition in the NFA from state A to state C while reading the symbol b. A production rule of the form $A \longrightarrow B$ corresponds to an ε-transition from state A to state B in the NFA. And a production rule of the form $A \longrightarrow \varepsilon$ exists in the grammar if and only if A is a final state in the NFA. With this correspondence, a derivation of a string w in the grammar corresponds to an execution path through the NFA as it accepts the string w. I won't give a complete proof here. You are welcome to work through the details if you want. But the important fact is:

Theorem 4.4. *A language L is regular if and only if there is a right-regular grammar G such that $L = L(G)$. In particular, every regular language is context-free.*

Exercises

1. Show that Part 4 of Theorem 4.1 follows from Part 3.
2. Give a careful proof that the language $\{a^n b^m \mid n \geq m \geq 0\}$ is generated by the context-free grammar

$$S \longrightarrow aSb$$
$$S \longrightarrow A$$
$$A \longrightarrow aA$$
$$A \longrightarrow \varepsilon$$

3. Identify the language generated by each of the following context-free grammars.

a) $S \longrightarrow aaSb$
$S \longrightarrow \varepsilon$

b) $S \longrightarrow aSb$
$S \longrightarrow aaSb$
$S \longrightarrow \varepsilon$

c) $S \longrightarrow TS$
$S \longrightarrow \varepsilon$
$T \longrightarrow aTb$
$T \longrightarrow \varepsilon$

d) $S \longrightarrow ABA$
$A \longrightarrow aA$
$A \longrightarrow a$
$B \longrightarrow bB$
$B \longrightarrow cB$
$B \longrightarrow \varepsilon$

4. For each of the following languages find a context-free grammar that generates the language:

a) $\{a^n b^m \mid n \geq m > 0\}$ **b)** $\{a^n b^m \mid n, m \in \mathbb{N}\}$
c) $\{a^n b^m \mid n \geq 0 \wedge m = n + 1\}$ **d)** $\{a^n b^m c^n \mid n, m \in \mathbb{N}\}$
e) $\{a^n b^m c^k \mid n = m + k\}$ **f)** $\{a^n b^m \mid n \neq m\}$
g) $\{a^n b^m c^r d^t \mid n + m = r + t\}$ **h)** $\{a^n b^m c^k \mid n \neq m + k\}$

5. Find a context-free grammar that generates the language $\{w \in \{a, b\}^* \mid n_a(w) > n_b(w)\}$.

6. Find a context-free grammar that generates the language $\{w \in \{a, b, c\}^* \mid n_a(w) = n_b(w)\}$.

7. A ***palindrome*** is a string that reads the same backwards and forwards, such as "mom", "radar", or "aabccbccbaa". That is, w is a palindrome if $w = w^R$. Let $L = \{w \in \{a, b, c\}^* \mid w \text{ is a palindrome }\}$. Show that L is a context-free language by finding a context-free grammar that generates L.

8. Let $\Sigma = \{$ `(`, `)`, `[`, `]` $\}$. That is, Σ is the alphabet consisting of the four symbols `(`, `)`, `[`, and `]`. Let L be the language over Σ consisting of strings in which both parentheses and brackets are balanced. For example, the string `([][()()])([])` is in L but `[(])` is not. Find a context-free grammar that generates the language L.

9. Suppose that G and H are context-free grammars. Let $L = L(G)$ and let $M = L(H)$. Explain how to construct a context-free grammar for the language LM. You do not need to give a formal proof that your grammar is correct.

10. Suppose that G is a context-free grammar. Let $L = L(G)$. Explain how to construct a context-free grammar for the language L^*. You do not need to give a formal proof that your grammar is correct.

11. Suppose that L is a context-free language. Prove that L^R is a context-free language. (Hint: Given a context-free grammar G for L, make a new grammar, G^R, by reversing the right-hand side of each of the production rules in G. That is, $A \longrightarrow w$ is a production rule in G if and only if $A \longrightarrow w^R$ is a production rule in G^R.)

12. Define a ***left-regular grammar*** to be a context-free grammar in which the right-hand side of every production rule is of one of the following forms: the empty string; a single non-terminal symbol; or a non-terminal symbol followed by a terminal symbol. Show that a language is regular if and only if it can be generated by a left-regular grammar. (Hint: Use the preceding exercise and Theorem 4.4.)

4.2 Application: BNF

Context-free grammars are used to describe some aspects of the syntax of programming languages. However, the notation that is used for grammars in the context of programming languages is somewhat different from the notation introduced in the preceding section. The notation that is used is called ***Backus-Naur Form*** or BNF. It is named after computer scientists John Backus and Peter Naur, who developed the notation. Actually, several variations of BNF exist. I will discuss one of them here. BNF can be used to describe the syntax of natural languages, as well as programming languages, and some of the examples in this section will deal with the syntax of English.

Like context-free grammars, BNF grammars make use of production rules, non-terminals, and terminals. The non-terminals are usually given meaningful, multi-character names. Here, I will follow a common practice of enclosing non-terminals in angle brackets, so that they can be easily distinguished. For example, $\langle noun \rangle$ and $\langle sentence \rangle$ could be non-terminals in a BNF grammar for English, while $\langle program \rangle$ and $\langle \textit{if-statement} \rangle$ might be used in a BNF grammar for a programming language. Note that a BNF non-terminal usually represents a meaningful ***syntactic category***, that is, a certain type of building block in the syntax of the language that is being described, such as an adverb, a prepositional phrase, or a variable declaration statement. The terminals of a BNF grammar are the things that actually appear in the language that is being described. In the case of natural language, the terminals are individual words.

In BNF production rules, I will use the symbol "::=" in place of the "$\longrightarrow$" that is used in context-free grammars. BNF production rules are more powerful than the production rules in context-free grammars. That is, one BNF rule might be equivalent to several context-free grammar rules. As for context-free grammars, the left-hand side of a BNF production rule is a single non-terminal symbol. The right hand side can include terminals and non-terminals, and can also use the following notations, which should remind you of notations used in regular expressions:

- A vertical bar, |, indicates a choice of alternatives. For example,

  $$\langle digit \rangle \ ::= 0 \mid 1 \mid 2 \mid 3 \mid 4 \mid 5 \mid 6 \mid 7 \mid 8 \mid 9$$

  indicates that the non-terminal $\langle digit \rangle$ can be replaced by any one of the terminal symbols 0, 1, ..., 9.

- Items enclosed in brackets are optional. For example,

  $$\langle \textit{declaration} \rangle ::= \langle \textit{type} \rangle\ \langle \textit{variable} \rangle\ [\ = \langle \textit{expression} \rangle\ ]\ ;$$

  says that $\langle \textit{declaration} \rangle$ can be replaced either by "$\langle \textit{type} \rangle\ \langle \textit{variable} \rangle$;" or by "$\langle \textit{type} \rangle\ \langle \textit{variable} \rangle = \langle \textit{expression} \rangle$;". (The symbols "=" and ";" are terminal symbols in this rule.)

- Items enclosed between "[" and "]..." can be repeated zero or more times. (This has the same effect as a "*"in a regular expression.) For example,

  $$\langle \textit{integer} \rangle ::= \langle \textit{digit} \rangle\ [\ \langle \textit{digit} \rangle\ ]\ldots$$

  says that an $\langle \textit{integer} \rangle$ consists of a $\langle \textit{digit} \rangle$ followed optionally by any number of additional $\langle \textit{digit} \rangle$'s. That is, the non-terminal $\langle \textit{integer} \rangle$ can be replaced by $\langle \textit{digit} \rangle$ or by $\langle \textit{digit} \rangle\langle \textit{digit} \rangle$ or by $\langle \textit{digit} \rangle\langle \textit{digit} \rangle\langle \textit{digit} \rangle$, and so on.

- Parentheses can be used as usual, for grouping.

All these notations can be expressed in a context-free grammar by introducing additional production rules. For example, the BNF rule "$\langle \textit{sign} \rangle ::= + \mid -$" is equivalent to the two rules, "$\langle \textit{sign} \rangle ::= +$" and "$\langle \textit{sign} \rangle ::= -$". A rule that contains an optional item can also be replaced by two rules. For example,

$$\langle \textit{declaration} \rangle ::= \langle \textit{type} \rangle\ \langle \textit{variable} \rangle\ [\ = \langle \textit{expression} \rangle\ ]\ ;$$

can be replaced by the two rules

$$\langle \textit{declaration} \rangle ::= \langle \textit{type} \rangle\ \langle \textit{variable} \rangle\ ;$$
$$\langle \textit{declaration} \rangle ::= \langle \textit{type} \rangle\ \langle \textit{variable} \rangle = \langle \textit{expression} \rangle\ ;$$

In context-free grammars, repetition can be expressed by using a recursive rule such as "$S \longrightarrow aS$", in which the same non-terminal symbol appears both on the left-hand side and on the right-hand side of the rule. BNF-style notation using "[" and "]..." can be eliminated by replacing it with a new non-terminal symbol and adding a recursive rule to allow that symbol to repeat zero or more times. For example, the production rule

$$\langle \textit{integer} \rangle ::= \langle \textit{digit} \rangle\ [\ \langle \textit{digit} \rangle\ ]\ldots$$

can be replaced by three rules using a new non-terminal symbol $\langle \textit{digit-list} \rangle$ to represent a string of zero or more $\langle \textit{digit} \rangle$'s:

$$\langle \textit{integer} \rangle ::= \langle \textit{digit} \rangle\ \langle \textit{digit-list} \rangle$$
$$\langle \textit{digit-list} \rangle ::= \langle \textit{digit} \rangle\ \langle \textit{digit-list} \rangle$$
$$\langle \textit{digit-list} \rangle ::= \varepsilon$$

As an example of a complete BNF grammar, let's look at a BNF grammar for a very small subset of English. The start symbol for the grammar is

⟨*sentence*⟩, and the terminal symbols are English words. All the sentences that can be produced from this grammar are syntactically correct English sentences, although you wouldn't encounter many of them in conversation. Here is the grammar:

> ⟨*sentence*⟩ ::= ⟨*simple-sentence*⟩ [and ⟨*simple-sentence*⟩]...
>
> ⟨*simple-sentence*⟩ ::= ⟨*nout-part*⟩ ⟨*verb-part*⟩
>
> ⟨*noun-part*⟩ ::= ⟨*article*⟩ ⟨*noun*⟩ [who ⟨*verb-part*⟩]...
>
> ⟨*verb-part*⟩ ::= ⟨*intransitive-verb*⟩ | (⟨*transitive-verb*⟩ ⟨*noun-part*⟩)
>
> ⟨*article*⟩ ::= the | a
>
> ⟨*noun*⟩ ::= man | woman | dog | cat | computer
>
> ⟨*intransitive-verb*⟩ ::= runs | jumps | hides
>
> ⟨*transitive-verb*⟩ ::= knows | loves | chases | owns

This grammar can generate sentences such as "A dog chases the cat and the cat hides" and "The man loves a woman who runs." The second sentence, for example, is generated by the derivation

$$\begin{aligned}
\langle \textit{sentence} \rangle &\Longrightarrow \langle \textit{simple-sentence} \rangle \\
&\Longrightarrow \langle \textit{noun-part} \rangle\ \langle \textit{verb-part} \rangle \\
&\Longrightarrow \langle \textit{article} \rangle\ \langle \textit{noun} \rangle\ \langle \textit{verb-part} \rangle \\
&\Longrightarrow \text{the } \langle \textit{noun} \rangle\ \langle \textit{verb-part} \rangle \\
&\Longrightarrow \text{the man } \langle \textit{verb-part} \rangle \\
&\Longrightarrow \text{the man } \langle \textit{transitive-verb} \rangle\ \langle \textit{noun-part} \rangle \\
&\Longrightarrow \text{the man loves } \langle \textit{noun-part} \rangle \\
&\Longrightarrow \text{the man loves } \langle \textit{article} \rangle\ \langle \textit{noun} \rangle \text{ who } \langle \textit{verb-part} \rangle \\
&\Longrightarrow \text{the man loves a } \langle \textit{noun} \rangle \text{ who } \langle \textit{verb-part} \rangle \\
&\Longrightarrow \text{the man loves a woman who } \langle \textit{verb-part} \rangle \\
&\Longrightarrow \text{the man loves a woman who } \langle \textit{intransitive-verb} \rangle \\
&\Longrightarrow \text{the man loves a woman who runs}
\end{aligned}$$

BNF is most often used to specify the syntax of programming languages. Most programming languages are not, in fact, context-free languages, and BNF is not capable of expressing all aspects of their syntax. For example, BNF cannot express the fact that a variable must be declared before it is used or the fact that the number of actual parameters in a subroutine call statement must match the number of formal parameters in the declaration of the subroutine. So BNF is used to express the context-free aspects of the syntax of a programming language, and other restrictions on the syntax—such as the rule about declaring a variable before it is used—are expressed using informal English descriptions.

When BNF is applied to programming languages, the terminal symbols are generally "tokens," which are the minimal meaningful units in a program. For example, the pair of symbols `<=` constitute a single token, as does a string such as `"Hello World"`. Every number is represented by a single token. (The actual value of the number is stored as a so-called "attribute" of the token, but the value plays no role in the context-free syntax of the language.) I will use the symbol ***number*** to represent a numerical token. Similarly, every variable name, subroutine name, or other identifier in the program is represented by the same token, which I will denote as ***ident***. One final complication: Some symbols used in programs, such as "]" and "(", are also used with a special meaning in BNF grammars. When such a symbol occurs as a terminal symbol, I will enclose it in double quotes. For example, in the BNF production rule

$$\langle \textit{array-reference} \rangle ::= \textbf{\textit{ident}} \text{ ``['' } \langle \textit{expression} \rangle \text{ ``]''}$$

the "[" and "]" are terminal symbols in the language that is being described, rather than the BNF notation for an optional item. With this notation, here is part of a BNF grammar that describes statements in the Java programming language:

$$\langle \textit{statement} \rangle ::= \langle \textit{block-statement} \rangle \mid \langle \textit{if-statement} \rangle \mid \langle \textit{while-statement} \rangle \\ \mid \langle \textit{assignment-statement} \rangle \mid \langle \textit{null-statement} \rangle$$

$$\langle \textit{block-statement} \rangle ::= \{ \; [\; \langle \textit{statement} \rangle \;]\dots \; \}$$

$$\langle \textit{if-statement} \rangle ::= \text{if ``('' } \langle \textit{condition} \rangle \text{ ``)'' } \langle \textit{statement} \rangle \; [\text{ else } \langle \textit{statement} \rangle \;]$$

$$\langle \textit{while-statement} \rangle ::= \text{while ``('' } \langle \textit{condition} \rangle \text{ ``)'' } \langle \textit{statement} \rangle$$

$$\langle \textit{assignment-statement} \rangle ::= \langle \textit{variable} \rangle = \langle \textit{expression} \rangle \; ;$$

$$\langle \textit{null-statement} \rangle ::= \varepsilon$$

The non-terminals $\langle \textit{condition} \rangle$, $\langle \textit{variable} \rangle$, and $\langle \textit{expression} \rangle$ would, of course, have to be defined by other production rules in the grammar. Here is a set of rules that define simple expressions, made up of numbers, identifiers, parentheses and the arithmetic operators $+$, $-$, $*$ and $/$:

$$\langle \textit{expression} \rangle ::= \langle \textit{term} \rangle \; [\; [+ \mid - \;] \; \langle \textit{term} \rangle \;]\dots$$

$$\langle \textit{term} \rangle ::= \langle \textit{factor} \rangle \; [\; [* \mid / \;] \; \langle \textit{factor} \rangle \;]\dots$$

$$\langle \textit{factor} \rangle ::= \textbf{\textit{ident}} \mid \textbf{\textit{number}} \mid \text{``('' } \langle \textit{expression} \rangle \text{ ``)''}$$

The first rule says that an $\langle \textit{expression} \rangle$ is a sequence of one or more $\langle \textit{term} \rangle$'s, separated by plus or minus signs. The second rule defines a $\langle \textit{term} \rangle$ to be a sequence of one or more $\langle factors \rangle$, separated by multiplication or division operators. The last rule says that a $\langle factor \rangle$ can be either an identifier or a number or an $\langle expression \rangle$ enclosed in parentheses. This small BNF grammar can generate expressions such as "$3*5$" and "$x*(x+1)-3/(z+2*(3-x))+7$". The latter expression is made up of three terms: $x*(x+1)$, $3/(z+2*(3-x))$, and 7. The first of these terms is made up of two factors, x and $(x+1)$. The factor $(x+1)$ consists of the expression $x+1$ inside a pair of parentheses.

The nice thing about this grammar is that the precedence rules for the operators are implicit in the grammar. For example, according to the grammar, the expression $3+5*7$ is seen as $\langle term \rangle + \langle term \rangle$ where the first term is 3 and the second term is $5*7$. The $5*7$ occurs as a group, which must be evaluated before the result is added to 3. Parentheses can change the order of evaluation. For example, $(3+5)*7$ is generated by the grammar as a single $\langle term \rangle$ of the form $\langle factor \rangle * \langle factor \rangle$. The first $\langle factor \rangle$ is $(3+5)$. When $(3+5)*7$ is evaluated, the value of $(3+5)$ is computed first and then multiplied by 7. This is an example of how a grammar that describes the syntax of a language can also reflect its meaning.

Although this section has not introduced any really new ideas or theoretical results, I hope it has demonstrated how context-free grammars can be applied in practice.

Exercises

1. One of the examples in this section was a grammar for a subset of English. Give five more examples of sentences that can be generated from that grammar. Your examples should, collectively, use all the rules of the grammar.

2. Rewrite the example BNF grammar for a subset of English as a context-free grammar.

3. Write a single BNF production rule that is equivalent to the following context-free grammar:

$$\begin{aligned} S &\longrightarrow aSa \\ S &\longrightarrow bB \\ B &\longrightarrow bB \\ B &\longrightarrow \varepsilon \end{aligned}$$

4. Write a BNF production rule that specifies the syntax of real numbers, as they appear in programming languages such as Java and C. Real numbers can include a sign, a decimal point and an exponential part. Some examples are: 17.3, .73, 23.1e67, −1.34E−12, +0.2, 100E+100

5. Variable references in the Java programming language can be rather complicated. Some examples include: x, $list.next$, $A[7]$, $a.b.c$, $S[i+1].grid[r][c].red$, ... Write a BNF production rule for Java variables. You can use the token ***ident*** and the non-terminal $\langle expression \rangle$ in your rule.
6. Use BNF to express the syntax of the try...catch statement in the Java programming language.
7. Give a BNF grammar for compound propositions made up of propositional variables, parentheses, and the logical operators $\wedge$, $\vee$, and $\neg$. Use the non-terminal symbol $\langle pv \rangle$ to represent a propositional variable. You do not have to give a definition of $\langle pv \rangle$.

4.3 Parsing and Parse Trees

Suppose that G is a grammar for the language L. That is, $L = L(G)$. The grammar G can be used to generate strings in the language L. In practice, though, we often start with a string which might or might not be in L, and the problem is to determine whether the string is in the language and, if so, how it can be generated by G. The goal is to find a derivation of the string, using the production rules of the grammar, or to show that no such derivation exists. This is known as ***parsing*** the string. When the string is a computer program or a sentence in a natural language, parsing the string is an essential step in determining its meaning.

As an example that we will use throughout this section, consider the language that consists of arithmetic expressions containing parentheses, the binary operators $+$ and $*$, and the variables x, y, and z. Strings in this language include x, $x + y * z$, and $((x + y) * y) + z * z$. Here is a context-free grammar that generates this language:

$$
\begin{aligned}
E &\longrightarrow E + E \\
E &\longrightarrow E * E \\
E &\longrightarrow (E) \\
E &\longrightarrow x \\
E &\longrightarrow y \\
E &\longrightarrow z
\end{aligned}
$$

Call the grammar described by these production rules G_1. The grammar G_1 says that x, y, and z are expressions, and that you can make new expressions by adding two expressions, by multiplying two expressions, and by enclosing an expression in parentheses. (Later, we'll look at other grammars for the same language—ones that turn out to have certain advantages over G_1.)

Consider the string $x+y*z$. To show that this string is in the language $L(G_1)$, we can exhibit a derivation of the string from the start symbol E. For example:

$$\begin{aligned} E &\Longrightarrow E+E \\ &\Longrightarrow E+E*E \\ &\Longrightarrow E+y*E \\ &\Longrightarrow x+y*E \\ &\Longrightarrow x+y*z \end{aligned}$$

This derivation shows that the string $x+y*z$ is in fact in $L(G_1)$. Now, this string has many other derivations. At each step in the derivation, there can be a lot of freedom about which rule in the grammar to apply next. Some of this freedom is clearly not very meaningful. When faced with the string $E+E*E$ in the above example, the order in which we replace the E's with the variables x, y, and z doesn't much matter. To cut out some of this meaningless freedom, we could agree that in each step of a derivation, the non-terminal symbol that is replaced is the leftmost non-terminal symbol in the string. A derivation in which this is true is called a ***left derivation***. The following left derivation of the string $x+y*z$ uses the same production rules as the previous derivation, but it applies them in a different order:

$$\begin{aligned} E &\Longrightarrow E+E \\ &\Longrightarrow x+E \\ &\Longrightarrow x+E*E \\ &\Longrightarrow x+y*E \\ &\Longrightarrow x+y*z \end{aligned}$$

It shouldn't be too hard to convince yourself that any string that has a derivation has a left derivation (which can be obtained by changing the order in which production rules are applied).

We have seen that the same string might have several different derivations. We might ask whether it can have several different left derivations. The answer is that it depends on the grammar. A context-free grammar G is said to be ***ambiguous*** if there is a string $w \in L(G)$ such that w has more than one left derivation according to the grammar G.

Our example grammar G_1 is ambiguous. In fact, in addition to the left derivation given above, the string $x+y*z$ has the alternative left derivation

$$\begin{aligned} E &\Longrightarrow E * E \\ &\Longrightarrow E + E * E \\ &\Longrightarrow x + E * E \\ &\Longrightarrow x + y * E \\ &\Longrightarrow x + y * z \end{aligned}$$

In this left derivation of the string $x+y*z$, the first production rule that is applied is $E \longrightarrow E*E$. The first E on the right-hand side eventually yields "$x + y$" while the second yields "z". In the previous left derivation, the first production rule that was applied was $E \longrightarrow E + E$, with the first E on the right yielding "x" and the second E yielding "$y * z$". If we think in terms of arithmetic expressions, the two left derivations lead to two different interpretations of the expression $x + y * z$. In one interpretation, the $x + y$ is a unit that is multiplied by z. In the second interpretation, the $y * z$ is a unit that is added to x. The second interpretation is the one that is correct according to the usual rules of arithmetic. However, the grammar allows either interpretation. The ambiguity of the grammar allows the string to be parsed in two essentially different ways, and only one of the parsings is consistent with the meaning of the string. Of course, the grammar for English is also ambiguous. In a famous example, it's impossible to tell whether a "pretty girls' camp" is meant to describe a pretty camp for girls or a camp for pretty girls.

When dealing with artificial languages such as programming languages, it's better to avoid ambiguity. The grammar G_1 is perfectly correct in that it generates the correct set of strings, but in a practical situation where we are interested in the meaning of the strings, G_1 is not the right grammar for the job. There are other grammars that generate the same language as G_1. Some of them are unambiguous grammars that better reflect the meaning of the strings in the language. For example, the language $L(G_1)$ is also generated by the BNF grammar

$$\begin{aligned} E &::= T\ [\ +\ T\ ]\ldots \\ T &::= F\ [\ *\ F\ ]\ldots \\ F &::= \text{"("}\ E\ \text{")"}\ \mid x \mid y \mid z \end{aligned}$$

This grammar can be translated into a standard context-free grammar, which I will call G_2:

$$
\begin{aligned}
E &\longrightarrow TA \\
A &\longrightarrow +TA \\
A &\longrightarrow \varepsilon \\
T &\longrightarrow FB \\
B &\longrightarrow *FB \\
B &\longrightarrow \varepsilon \\
F &\longrightarrow (E) \\
F &\longrightarrow x \\
F &\longrightarrow y \\
F &\longrightarrow z
\end{aligned}
$$

The language generated by G_2 consists of all legal arithmetic expressions made up of parentheses, the operators $+$ and $-$, and the variables x, y, and z. That is, $L(G_2) = L(G_1)$. However, G_2 is an unambiguous grammar. Consider, for example, the string $x+y*z$. Using the grammar G_2, the only left derivation for this string is:

$$
\begin{aligned}
E &\Longrightarrow TA \\
&\Longrightarrow FBA \\
&\Longrightarrow xBA \\
&\Longrightarrow xA \\
&\Longrightarrow x + TA \\
&\Longrightarrow x + FBA \\
&\Longrightarrow x + yBA \\
&\Longrightarrow x + y * FBA \\
&\Longrightarrow x + y * zBA \\
&\Longrightarrow x + y * zA \\
&\Longrightarrow x + y * z
\end{aligned}
$$

There is no choice about the first step in this derivation, since the only production rule with E on the left-hand side is $E \longrightarrow TA$. Similarly, the second step is forced by the fact that there is only one rule for rewriting a T. In the third step, we must replace an F. There are four ways to rewrite F, but only one way to produce the x that begins the string $x + y * z$, so we apply the rule $F \longrightarrow x$. Now, we have to decide what to do with the B in xBA. There two rules for rewriting B, $B \longrightarrow *FB$ and $B \longrightarrow \varepsilon$.

However, the first of these rules introduces a non-terminal, $*$, which does not match the string we are trying to parse. So, the only choice is to apply the production rule $B \longrightarrow \varepsilon$. In the next step of the derivation, we must apply the rule $A \longrightarrow +TA$ in order to account for the $+$ in the string $x + y * z$. Similarly, each of the remaining steps in the left derivation is forced.

The fact that G_2 is an unambiguous grammar means that at each step in a left derivation for a string w, there is only one production rule that can be applied which will lead ultimately to a correct derivation of w. However, G_2 actually satisfies a much stronger property: at each step in the left derivation of w, we can tell which production rule has to be applied by looking ahead at the next symbol in w. We say that G_2 is an ***LL(1) grammar***. (This notation means that we can read a string from **L**eft to right and construct a **L**eft derivation of the string by looking ahead at most **1** character in the string.) Given an LL(1) grammar for a language, it is fairly straightforward to write a computer program that can parse strings in that language. If the language is a programming language, then parsing is one of the essential steps in translating a computer program into machine language. LL(1) grammars and parsing programs that use them are often studied in courses in programming languages and the theory of compilers.

Not every unambiguous context-free grammar is an LL(1) grammar. Consider, for example, the following grammar, which I will call G_3:

$$
\begin{aligned}
&E \longrightarrow E + T\\
&E \longrightarrow T\\
&T \longrightarrow T * F\\
&T \longrightarrow F\\
&F \longrightarrow (E)\\
&F \longrightarrow x\\
&F \longrightarrow y\\
&F \longrightarrow z
\end{aligned}
$$

This grammar generates the same language as G_1 and G_2, and it is unambiguous. However, it is not possible to construct a left derivation for a string according to the grammar G_3 by looking ahead one character in the string at each step. The first step in any left derivation must be either $E \Longrightarrow E + T$ or $E \Longrightarrow T$. But how can we decide which of these is the correct first step? Consider the strings $(x + y) * z$ and $(x + y) * z + z * x$, which are both in the language $L(G_3)$. For the string $(x + y) * z$, the first step in a left derivation must be $E \Longrightarrow T$, while the first step in a left derivation of $(x+y)*z+z*x$ must be $E \Longrightarrow E+T$. However, the first seven characters of the strings are identical, so clearly looking even seven characters ahead is not enough to tell us which production rule to apply. In fact, similar examples show that looking ahead any given finite number of characters is not enough.

However, there is an alternative parsing procedure that will work for G_3. This alternative method of parsing a string produces a ***right derivation*** of the string, that is, a derivation in which at each step, the non-terminal symbol that is replaced is the rightmost non-terminal symbol in the string. Here, for example, is a right derivation of the string $(x+y)*z$ according to the grammar G_3:

$$\begin{aligned} E &\Longrightarrow T \\ &\Longrightarrow T*F \\ &\Longrightarrow T*z \\ &\Longrightarrow F*z \\ &\Longrightarrow (E)*z \\ &\Longrightarrow (E+T)*z \\ &\Longrightarrow (E+F)*z \\ &\Longrightarrow (E+y)*z \\ &\Longrightarrow (T+y)*z \\ &\Longrightarrow (F+y)*z \\ &\Longrightarrow (x+y)*z \end{aligned}$$

The parsing method that produces this right derivation produces it from "bottom to top." That is, it begins with the string $(x+y)*z$ and works backward to the start symbol E, generating the steps of the right derivation in reverse order. The method works because G_3 is what is called an ***LR(1) grammar***. That is, roughly, it is possible to read a string from **L**eft to right and produce a **R**ight derivation of the string, by looking ahead at most **1** symbol at each step. Although LL(1) grammars are easier for people to work with, LR(1) grammars turn out to be very suitable for machine processing, and they are used as the basis for the parsing process in many compilers.

LR(1) parsing uses a ***shift/reduce*** algorithm. Imagine a cursor or current position that moves through the string that is being parsed. We can visualize the cursor as a vertical bar, so for the string $(x+y)*z$, we start with the configuration $|(x+y)*z$. A *shift* operation simply moves the cursor one symbol to the right. For example, a shift operation would convert $|(x + y) * z$ to $(|x + y) * z$, and a second shift operation would convert that to $(x| + y) * z$. In a reduce operation, one or more symbols immediately to the left of the cursor are recognized as the right-hand side of one of the production rules in the grammar. These symbols are removed and replaced by the left-hand side of the production rule. For example, in the configuration $(x| + y) * z$, the x to the left of the cursor is the right-hand side of the production rule $F \longrightarrow x$, so we can apply a reduce operation and replace the x with F, giving $(F| + y) * z$. This first reduce operation corresponds to the last step in the right derivation of the string, $(F + y) * z \Longrightarrow (x + y) * z$. Now the F can be recognized as the right-hand side of the production rule $T \longrightarrow F$, so we can replace the F with T, giving $(T|+y)*z$. This corresponds to the next-to-last step in the right derivation, $(T + y) * z \Longrightarrow (F + y) * z$.

At this point, we have the configuration $(T| + y) * z$. The T could be the right-hand side of the production rule $E \longrightarrow T$. However, it could also conceivably come from the rule $T \longrightarrow T * F$. How do we know whether to reduce the T to E at this point or to wait for a $*F$ to come along so that we can reduce $T * F$? We can decide by looking ahead at the next character after the cursor. Since this character is a $+$ rather than a $*$, we should choose the reduce operation that replaces T with E, giving $(E| + y) * z$. What makes G_3 an LR(1) grammar is the fact that we can always decide what operation to apply by looking ahead at most one symbol past the cursor.

After a few more shift and reduce operations, the configuration becomes $(E)| * z$, which we can reduce to $T| * z$ by applying the production rules $F \longrightarrow (E)$ and $T \longrightarrow F$. Now, faced with $T| * z$, we must once again decide between a shift operation and a reduce operation that applies the rule $E \longrightarrow T$. In this case, since the next character is a $*$ rather than a $+$, we apply the shift operation, giving $T*|z$. From there we get, in succession, $T * z|$, $T * F|$, $T|$, and finally $E|$. At this point, we have reduced the entire string $(x + y) * z$ to the start symbol of the grammar. The very last step, the reduction of T to E corresponds to the first step of the right derivation, $E \Longrightarrow T$.

In summary, LR(1) parsing transforms a string into the start symbol of the grammar by a sequence of shift and reduce operations. Each reduce operation corresponds to a step in a right derivation of the string, and these steps are generated in reverse order. Because the steps in the derivation are generated from "bottom to top," LR(1) parsing is a type of ***bottom-up parsing***. LL(1) parsing, on the other hand, generates the steps in a left derivation from "top to bottom" and so is a type of ***top-down parsing***.

Although the language generated by a context-free grammar is defined in terms of derivations, there is another way of presenting the generation of a string that is often more useful. A ***parse tree*** displays the generation of a string from the start symbol of a grammar as a two dimensional diagram. Here are two parse trees that show two derivations of the string x+y*z according to the grammar G_1, which was given at the beginning of this section:

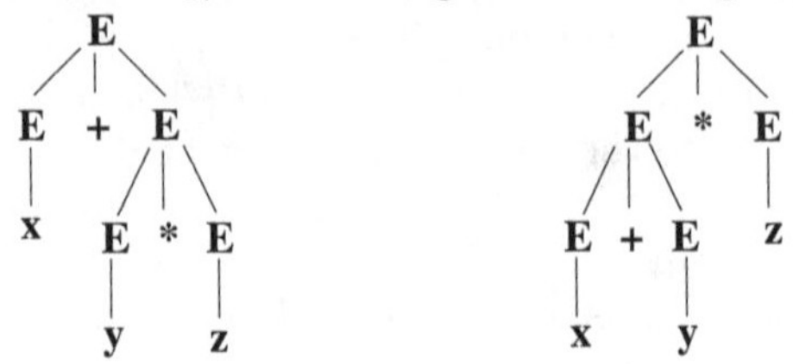

A parse tree is made up of terminal and non-terminal symbols, connected by lines. The start symbol is at the top, or "root," of the tree. Terminal symbols are at the lowest level, or "leaves," of the tree. (For some reason, computer scientists traditionally draw trees with leaves at the bottom and root at the top.) A production rule $A \longrightarrow w$ is represented in a parse tree by the symbol A lying above all the symbols in w, with a line joining A to each of the symbols in w. For example, in the left parse tree above, the root, E, is connected to the symbols E, $+$, and E, and this corresponds to an application of the production rule $E \longrightarrow E + E$.

It is customary to draw a parse tree with the string of non-terminals in a row across the bottom, and with the rest of the tree built on top of that base. Thus, the two parse trees shown above might be drawn as:

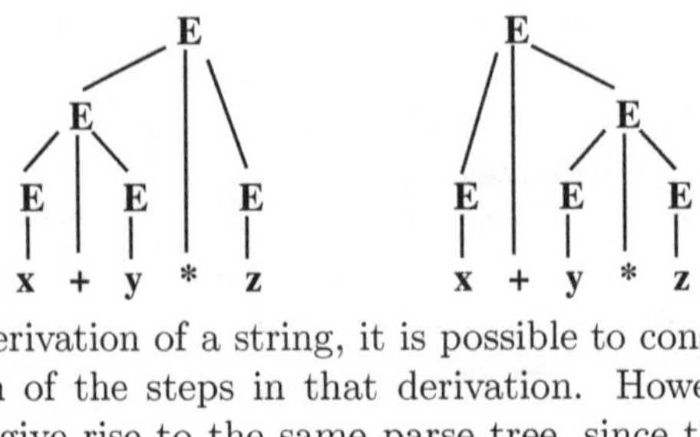

Given any derivation of a string, it is possible to construct a parse tree that shows each of the steps in that derivation. However, two different derivations can give rise to the same parse tree, since the parse tree does not show the order in which production rules are applied. For example, the parse tree on the left, above, does not show whether the production rule $E \longrightarrow x$ is applied before or after the production rule $E \longrightarrow y$. However, if we restrict our attention to left derivations, then we find that each parse tree corresponds to a unique left derivation and *vice versa*. I will state this fact as a theorem, without proof. A similar result holds for right derivations.

Theorem 4.5. *Let G be a context-free grammar. There is a one-to-one correspondence between parse trees and left derivations based on the grammar G.*

Based on this theorem, we can say that a context-free grammar G is ambiguous if and only if there is a string $w \in L(G)$ which has two parse trees.

Exercises

1. Show that each of the following grammars is ambiguous by finding a string that has two left derivations according to the grammar:

a) $S \longrightarrow SS$	**b)** $S \longrightarrow ASb$
$S \longrightarrow aSb$	$S \longrightarrow \varepsilon$
$S \longrightarrow bSa$	$A \longrightarrow aA$
$S \longrightarrow \varepsilon$	$A \longrightarrow a$

2. Consider the string $z{+}(x{+}y){*}x$. Find a left derivation of this string according to each of the grammars G_1, G_2, and G_3, as given in this section.
3. Draw a parse tree for the string $(x{+}y){*}z{*}x$ according to each of the grammars G_1, G_2, and G_3, as given in this section.
4. Draw three different parse trees for the string *ababbaab* based on the grammar given in part a) of exercise 1.
5. Suppose that the string *abbcabac* has the following parse tree, according to some grammar G:

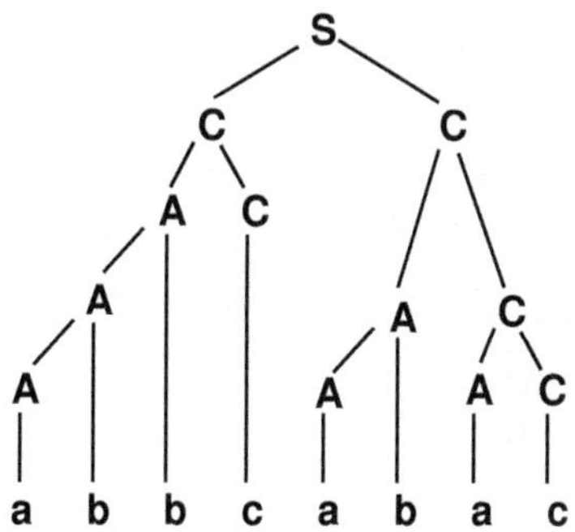

   a) List five production rules that must be rules in the grammar G, given that this is a valid parse tree.
   b) Give a left derivation for the string *abbcabac* according to the grammar G.
   c) Give a right derivation for the string *abbcabac* according to the grammar G.

6. Show the full sequence of shift and reduce operations that are used in the LR(1) parsing of the string $x + (y) * z$ according to the grammar G_3, and give the corresponding right derivation of the string.

7. This section showed how to use LL(1) and LR(1) parsing to find a derivation of a string in the language $L(G)$ generated by some grammar G. How is it possible to use LL(1) or LR(1) parsing to determine for an arbitrary string w whether $w \in L(G)$? Give an example.

4.4 Pushdown Automata

In the previous chapter, we saw that there is a neat correspondence between regular expressions and finite automata. That is, a language is generated by a regular expression if and only if that language is accepted by a finite automaton. Finite automata come in two types, deterministic and nondeterministic, but the two types of finite automata are equivalent in terms of their ability to recognize languages. So, the class of regular languages can be defined in two ways: either as the set of languages that can be generated by regular expressions or as the set of languages that can be recognized by finite automata (either deterministic or nondeterministic).

In this chapter, we have introduced the class of context-free languages, and we have considered how context-free grammars can be used to generate context-free languages. You might wonder whether there is any type of automaton that can be used to recognize context-free languages. In fact, there is: The abstract machines known as ***pushdown automata*** can be used to define context-free languages. That is, a language is context-free if and only if there is a pushdown automaton that accepts that language.

A pushdown automaton is essentially a finite automaton with an auxiliary data structure known as a ***stack***. A stack consists of a finite list of symbols. Symbols can be added to and removed from the list, but only at one end of the list. The end of the list where items can be added and removed is called the ***top*** of the stack. The list is usually visualized as a vertical "stack" of symbols, with items being added and removed at the top. Adding a symbol at the top of the stack is referred to as ***pushing*** a symbol onto the stack, and removing a symbol is referred to as ***popping*** an item from the stack. During each step of its computation, a pushdown automaton is capable of doing several push and pop operations on its stack (this in addition to possibly reading a symbol from the input string that is being processed by the automaton).

Before giving a formal definition of pushdown automata, we will look at how they can be represented by transition diagrams. A diagram of a pushdown automaton is similar to a diagram for an NFA, except that each transition in the diagram can involve stack operations. We will use a label of the form $\sigma, x/y$ on a transition to mean that the automaton consumes σ from its input string, pops x from the stack, and pushes y onto the stack. σ can be either ε or a single symbol. x and y are strings, possibly empty. (When a string $x = a_1a_2 \dots a_k$ is pushed onto a stack, the symbols are pushed in the order $a_k, \dots, a_1$, so that a_1 ends up on the top of the stack; for $y = b_1b_2 \dots b_n$ to be popped from the stack, b_1 must be the top symbol on the stack, followed by b_2, etc.) For example, consider the following transition diagram for a pushdown automaton:

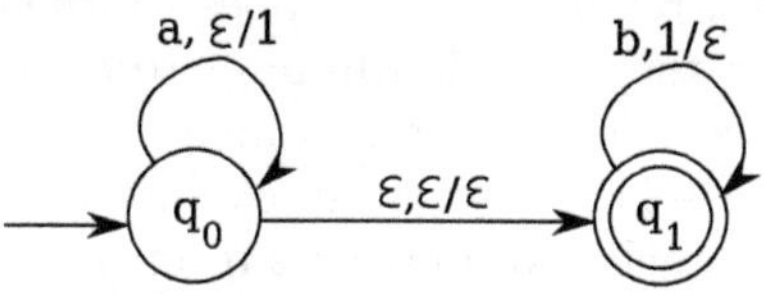

This pushdown automaton has start state q_0 and one accepting state, q_1. It can read strings over the alphabet $\Sigma = \{a, b\}$. The transition from q_0 to q_0, labeled with $a, \varepsilon/1$, means that if the machine is in state q_0, then it can read an a from its input string, pop nothing from the stack, push 1 onto the stack, and remain in state q_0. Similarly, the transition from q_1 to q_1 means that if the machine is in state q_1, it can read a b from its input string, pop a 1 from the stack, and push nothing onto the stack. Finally, the transition from state q_0 to q_1, labeled with $\varepsilon, \varepsilon/\varepsilon$, means that the machine can transition from state q_0 to state q_1 without reading, pushing, or popping anything.

Note that the automation can follow transition $b, 1/\varepsilon$ only if the next symbol in the input string is b and if 1 is on the top of the stack. When it makes the transition, it consumes the b from input and pops the 1 from the stack. Since in this case, the automaton pushes ε (that is, no symbols at all) onto the stack, the net change in the stack is simply to pop the 1.

We have to say what it means for this pushdown automaton to accept a string. For $w \in \{a, b\}^*$, we say that the pushdown automaton accepts w if and only if it is possible for the machine to start in its start state, q_0, read all of w, and finish in the accepting state, q_1, with an empty stack. Note in particular that it is not enough for the machine to finish in an accepting state—it must also empty the stack.[1]

It's not difficult to see that with this definition, the language accepted

[1]We could relax this restriction and require only that the machine finish in an accepting state after reading the string w, without requiring that the stack be empty. In fact, using this definition of accepting would not change the class of languages that are accepted by pushdown automata.

by our pushdown automaton is $\{a^n b^n \mid n \in \mathbb{N}\}$. In fact, given the string $w = a^k b^k$, the machine can process this string by following the transition from q_0 to q_0 k times. This will consume all the a's and will push k 1's onto the stack. The machine can then jump to state q_1 and follow the transition from q_1 to q_1 k times. Each time it does so, it consumes one b from the input and pops one 1 from the stack. At the end, the input has been completely consumed and the stack is empty. So, the string w is accepted by the automaton. Conversely, this pushdown automaton *only* accepts strings of the form $a^k b^k$, since the only way that the automaton can finish in the accepting state, q_1, is to follow the transition from q_0 to q_0 some number of times, reading a's as it does so, then jump at some point to q_1, and then follow the transition from q_1 to q_1 some number of times, reading b's as it does so. This means that an accepted string must be of the form $a^k b^\ell$ for some $k, \ell \in \mathbb{N}$. However, in reading this string, the automaton pushes k 1's onto the stack and pops ℓ 1's from the stack. For the stack to end up empty, ℓ must equal k, which means that in fact the string is of the form $a^k b^k$, as claimed.

Here are two more examples. These pushdown automata use the capability to push or pop more than one symbol at a time:

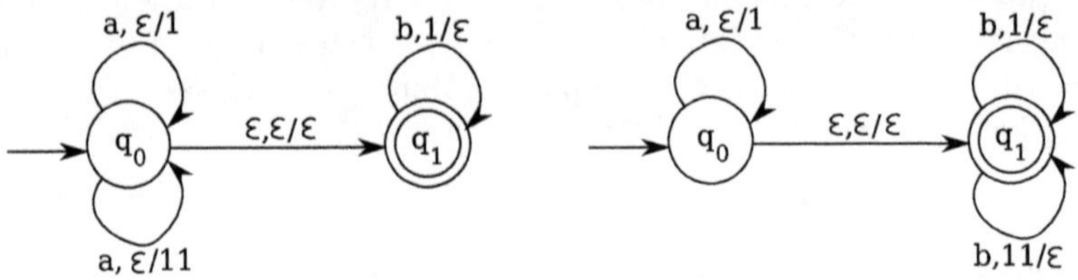

The atomaton on the left accepts the language $\{a^n b^m \mid n \leq m \leq 2 * n\}$. Each time it reads an a, it pushes either one or two 1's onto the stack, so that after reading n a's, the number of 1's on the stack is between n and $2*n$. If the machine then jumps to state q_1, it must be able to read exactly enough b's to empty the stack, so any string accepted by this machine must be of the form $a^n b^m$ with $n \leq m \leq 2 * n$. Conversely, any such string can be accepted by the machine. Similarly, the automaton on the right above accepts the language $\{a^n b^m \mid n/2 \leq m \leq n\}$. To accept $a^n b^m$, it must push n 1'a onto the stack and then pop one or two 1's for each b; this can succeed only if the number of b's is between $n/2$ and n.

Note that an NFA can be considered to be a pushdown automaton that does not make any use of its stack. This means that any language that can be accepted by an NFA (that is, any regular language) can be accepted by a pushdown automaton. Since the language $\{a^n b^n \mid n \in \mathbb{N}\}$ is context-free but not regular, and since it is accepted by the above pushdown automaton, we see that pushdown automata are capable of recognizing context-free languages that are not regular and so that pushdown automata are strictly more powerful than finite automata.

Although it is not particularly illuminating, we can give a formal definition of pushdown automaton. The definition does at least make it clear that the set of symbols that can be used on the stack is not necessarily the same as the set of symbols that can be used as input.

Definition 4.4. A pushdown automaton M is specified by six components $M = (Q, \Sigma, \Lambda, q_0, \partial, F)$ where

- Q is a finite set of states.
- Σ is an alphabet. Σ is the ***input alphabet*** for M.
- Λ is an alphabet. Λ is the ***stack alphabet*** for M.
- $q_0 \in Q$ is the ***start state*** of M.
- $F \subseteq Q$ is the set of ***final*** or ***accepting*** states in M.
- ∂ is the set of transitions in M. ∂ can be taken to be a finite subset of the set $\big(Q \times (\Sigma \cup \{\varepsilon\}) \times \Lambda^*\big) \times \big(Q \times \Lambda^*\big)$. An element $\big((q_1, \sigma, x), (q_2, y)\big)$ of ∂ represents a transition from state q_1 to state q_2 in which M reads σ from its input string, pops x from the stack, and pushes y onto the stack.

We can then define the language $L(M)$ accepted by a pushdown automaton $M = (Q, \Sigma, \Lambda, q_0, \partial, F)$ to be the set $L(M) = \{w \in \Sigma^* \mid$ starting from state q_0, it is possible for M to read all of w and finish in some state in F with an empty stack$\}$. With this definition, the class of languages accepted by pushdown automata is the same as the class of languages generated by context-free grammars.

Theorem 4.6. *Let Σ be an alphabet, and let L be a language over L. Then L is context-free if and only if there is a pushdown automaton whose input alphabet is Σ such that $L = L(M)$.*

We will not prove this theorem, but we do discuss how one direction can be proved. Suppose that L is a context-free language over an alphabet Σ. Let $G = (V, \Sigma, P, S)$ be a context-free grammar for L. Then we can construct a pushdown automaton M that accepts L. In fact, we can take $M = (Q, \Sigma, \Lambda, q_0, \partial, F)$ where $Q = \{q_0, q_1\}$, $\Lambda = \Sigma \cup V$, $F = \{q_1\}$, and ∂ contains transitions of the forms

1. $\big((q_0, \varepsilon, \varepsilon), (q_1, S)\big)$;

2. $\big((q_1, \sigma, \sigma), (q_1, \varepsilon)\big)$, for $\sigma \in \Sigma$; and

3. $\big((q_1, \varepsilon, A), (q_1, x)\big)$, for each production $A \longrightarrow x$ in G.

The transition $\big((q_0, \varepsilon, \varepsilon), (q_1, S)\big)$ lets M move from the start state q_0 to the accepting state q_1 while reading no input and pushing S onto the stack. This is the only possible first move by M.

A transition of the form $\big((q_1, \sigma, \sigma), (q_1, \varepsilon)\big)$, for $\sigma \in \Sigma$ allows M to read σ from its input string, provided there is a σ on the top of the stack. Note that if σ is at the top of the stack, then this transition is *only* transition that applies. Effectively, any terminal symbol that appears at the top of the stack must be matched by the same symbol in the input string, and the transition rule allows M to consume the symbol from the input string and remove it from the stack at the same time.

A transition of the third form, $\big((q_1, \varepsilon, A), (q_1, x)\big)$ can be applied if and only if the non-terminal symbol A is at the top of the stack. M consumes no input when this rule is applied, but A is replaced on the top of the stack by the string on the right-hand side of the production rule $A \longrightarrow x$. Since the grammar G can contain several production rules that have A as their left-hand side, there can be several transition rules in M that apply when A is on the top of the stack. This is the only source of nondeterminism in M; note that is also the source of nondeterminism in G.

The proof that $L(M) = L(G)$ follows from the fact that a computation of M that accepts a string $w \in \Sigma^*$ corresponds in a natural way to a left derivation of w from G's start symbol, S. Instead of giving a proof of this fact, we look at an example. Consider the following context-free grammar:

$$
\begin{aligned}
S &\longrightarrow AB \\
A &\longrightarrow aAb \\
A &\longrightarrow \varepsilon \\
B &\longrightarrow bB \\
B &\longrightarrow b
\end{aligned}
$$

This grammar generates the language $\{a^n b^m \mid m > n\}$. The pushdown automaton constructed from this grammar by the procedure given above has the following set of transition rules:

$$((q_0, \varepsilon, \varepsilon), (q_1, S))$$
$$((q_1, a, a), (q_1, \varepsilon))$$
$$((q_1, b, b), (q_1, \varepsilon))$$
$$((q_1, \varepsilon, S), (q_1, AB))$$
$$((q_1, \varepsilon, A), (q_1, aAb))$$
$$((q_1, \varepsilon, A), (q_1, \varepsilon))$$
$$((q_1, \varepsilon, B), (q_1, bB))$$
$$((q_1, \varepsilon, B), (q_1, b))$$

Suppose that the automaton is run on the input *aabbbb*. We can trace the sequence of transitions that are applied in a computation that accepts this input, and we can compare that computation to a left derivation of the string:

Transition	Input Consumed	Stack	Derivation
$((q_0, \varepsilon, \varepsilon), (q_1, S))$		S	
$((q_1, \varepsilon, S), (q_1, AB))$		AB	$S \Longrightarrow AB$
$((q_1, \varepsilon, A), (q_1, aAb))$		$aAbB$	$\Longrightarrow aAbB$
$((q_1, a, a), (q_1, \varepsilon))$	a	AbB	
$((q_1, \varepsilon, A), (q_1, aAb))$	a	$aAbbB$	$\Longrightarrow aaAbbB$
$((q_1, a, a), (q_1, \varepsilon))$	aa	$AbbB$	
$((q_1, \varepsilon, A), (q_1, \varepsilon))$	aa	bbB	$\Longrightarrow aabbB$
$((q_1, b, b), (q_1, \varepsilon))$	aab	bB	
$((q_1, b, b), (q_1, \varepsilon))$	$aabb$	B	
$((q_1, \varepsilon, B), (q_1, bB))$	$aabb$	bB	$\Longrightarrow aabbbB$
$((q_1, b), (q_1, b, \varepsilon))$	$aabbb$	B	
$((q_1, \varepsilon, B), (q_1, b))$	$aabbb$	b	$\Longrightarrow aabbbb$
$((q_1, b, b), (q_1, \varepsilon))$	$aabbbb$		

Note that at all times during this computation, the concatenation of the input that has been consumed so far with the contents of the stack is equal to one of the strings in the left derivation. Application of a rule of the form $((q_1, \sigma, \sigma), (q_1, \varepsilon))$ has the effect of removing one terminal symbol from the "Stack" column to the "Input Consumed" column. Application of a rule of the form $((q_1, \varepsilon, A), (q_1, x))$ has the effect of applying the next step in the left derivation to the non-terminal symbol on the top of the stack. (In the "Stack" column, the pushdown automaton's stack is shown with its top on the left.) In the end, the entire input string has been consumed and the stack is empty, which means that the string has been accepted by the pushdown automaton. It should be easy to see that for any context free grammar G, the same correspondence will always hold between left derivations and computations performed by the pushdown automaton constructed from G.

The computation of a pushdown automaton can involve nondeterminism. That is, at some point in the computation, there might be more than one transition rule that apply. When this is not the case—that is, when there is no circumstance in which two different transition rules apply—then we say that the pushdown automaton is ***deterministic***. Note that a deterministic pushdown automaton can have transition rules of the form $((q_i, \varepsilon, x), (q_j, y))$ (or even $((q_i, \varepsilon, \varepsilon), (q_j, y))$ if that is the *only* transition from state q_i). Note also that is is possible for a deterministic pushdown automaton to get "stuck"; that is, it is possible that no rules apply in some circumstances even though the input has not been completely consumed or the stack is not empty. If a deterministic pushdown automaton gets stuck while reading a string x, then x is not accepted by the automaton.

The automaton given at the beginning of this section, which accepts the language $\{a^n b^n \mid n \in \mathbb{N}\}$, is not deterministic. However, it is easy to construct a deterministic pushdown automaton for this language:

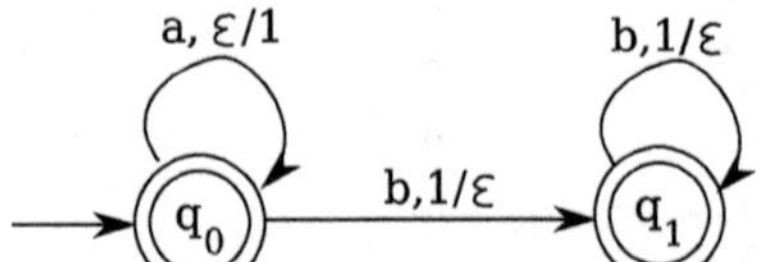

However, consider the language $\{ww^R \mid w \in \{a, b\}^*\}$. Here is a pushdown automaton that accepts this language:

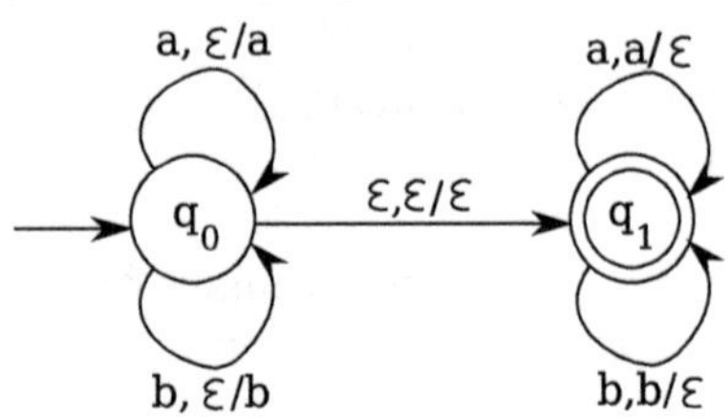

In state q_0, this machine copies the first part of its input string onto the stack. In state q_1, it tries to match the remainder of the input against the contents of the stack. In order for this to work, it must "guess" where the middle of the string occurs by following the transition from state q_0 to state q_1. In this case, it is by no means clear that it is possible to construct a deterministic pushdown automaton that accepts the same language.

At this point, it might be tempting to define a deterministic context-free language as one for which there exists a deterministic pushdown automaton which accepts that language. However, there is a technical problem with this definition: we need to make it possible for the pushdown automaton to detect the end of the input string. Consider the language $\{w \mid w \in \{a,b\}^* \wedge n_a(w) = n_b(w)\}$, which consists of strings over the alphabet $\{a,b\}$ in which the number of a's is equal to the number of b's. This language is accepted by the following pushdown automaton:

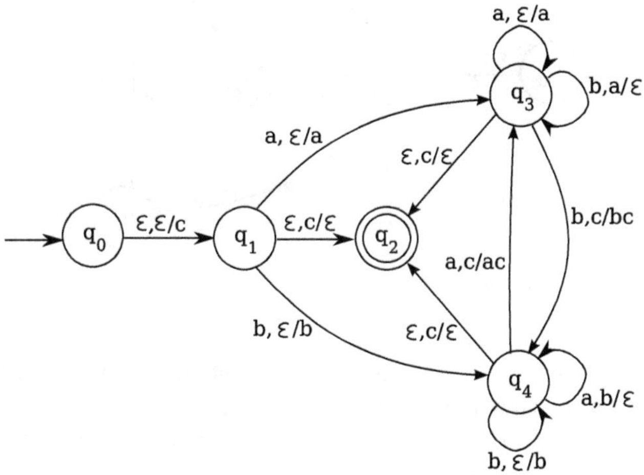

In this automaton, a c is first pushed onto the stack, and it remains on the bottom of the stack until the computation ends. During the process of reading an input string, if the machine is in state q_3, then the number of a's that have been read is greater than or equal to the number of b's that have been read, and the stack contains (copies of) the excess a's that have been read. Similarly, if the machine is in state q_4, then the number of b's that have been read is greater than or equal to the number of a's that have been read, and the stack contains (copies of) the excess b's that have been read. As the computation proceeds, if the stack contains nothing but a c, then the number of a's that have been consumed by the machine is equal to the number of b's that have been consumed; in such cases, the machine can pop the c from the stack—leaving the stack empty—and jump to state q_2. If the entire string has been read at that time, then the string is accepted. This involves nondeterminism because the automaton has to "guess" when to jump to state q_2; it has no way of knowing whether it has actually reached the end of the string.

Although this pushdown automaton is not deterministic, we can modify it easily to get a deterministic pushdown automaton that accepts a closely related language. We just have to add a special end-of-string symbol to the language. We use the symbol \$ for this purpose. The following deterministic automaton accepts the language $\{w\$ \mid w \in \{a,b\}^* \wedge n_a(w) = n_b(w)\}$:

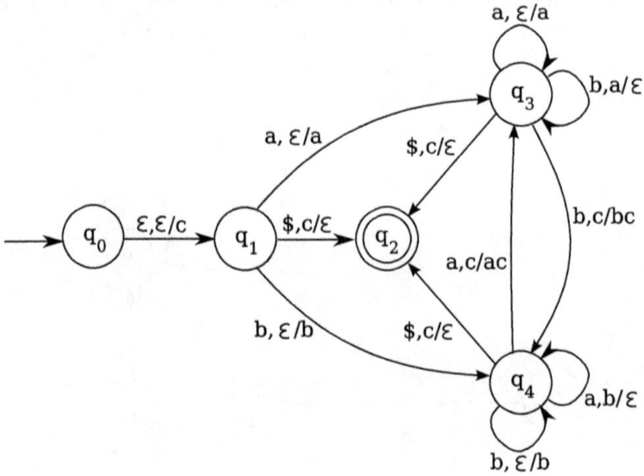

In this modified automaton, it is only possible for the machine to reach the accepting state q_2 by reading the end-of-string symbol at a time when the number of a's that have been consumed is equal to the number of b's. Taking our cue from this example, we define what it means for a language to be deterministic context-free as follows:

Definition 4.5. Let L be a language over an alphabet Σ, and let \$ be a symbol that is not in Σ. We say that L is a ***deterministic context-free language*** if there is a deterministic pushdown automaton that accepts the language $L\$$ (which is equal to $\{w\$ \mid w \in L\}$).

There are context-free languages that are not deterministic context-free. This means that for pushdown automata, nondeterminism adds real power. This contrasts with the case of finite automata, where deterministic finite automata and nondeterministic finite automata are equivalent in power in the sense that they accept the same class of languages.

A deterministic context-free language can be parsed efficiently. LL(1) parsing and LR(1) parsing can both be defined in terms of deterministic pushdown automata, although we have not pursued that approach here.

Exercises

1. Identify the context-free language that is accepted by each of the following pushdown automata. Explain your answers.

a)

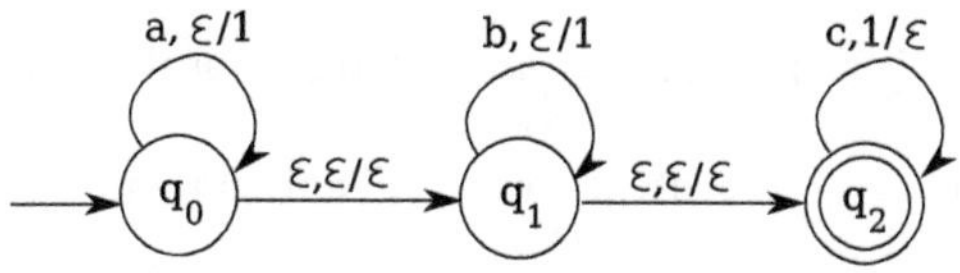

b)

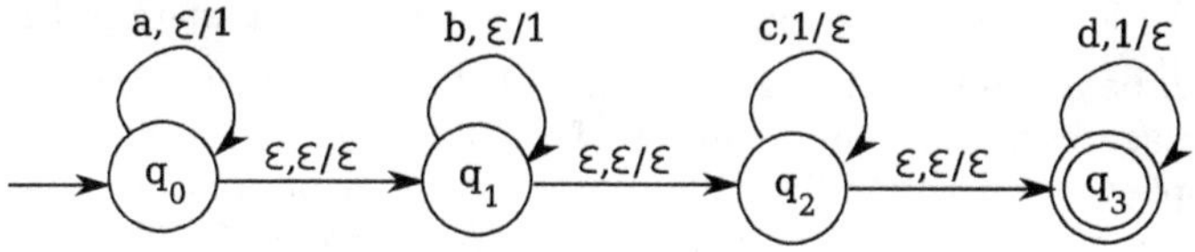

c)

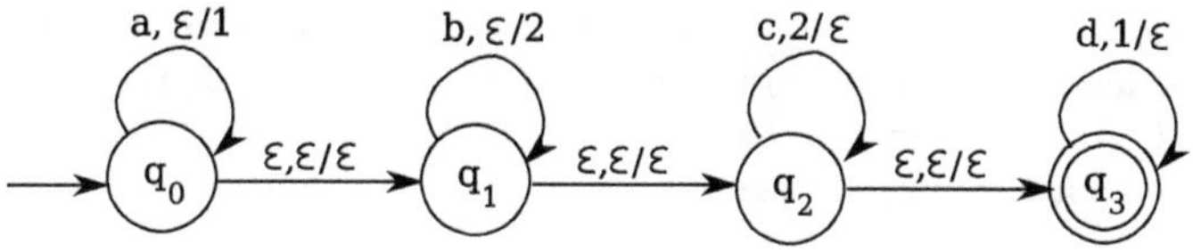

d)

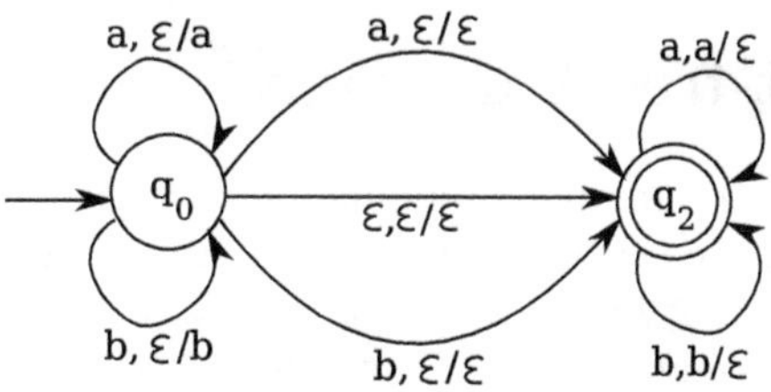

2. Let B be the language over the alphabet $\{\, (\, ,\,)\, \}$ that consists of strings of parentheses that are balanced in the sense that every left parenthesis has a matching right parenthesis. Examples include (), (())(), ((())())()(()), and the empty string. Find a deterministic pushdown automaton with a single state that accepts the language B. Explain how your automaton works, and explain the circumstances in which it will *fail* to accept a given string of parentheses.

3. Suppose that L is language over an alphabet Σ. Suppose that there is a deterministic pushdown automaton that accepts L. Show that L is deterministic context-free. That is, show how to construct a deterministic pushdown automaton that accepts the language $L\$$. (Assume that the symbol \$ is not in Σ.)

4. Find a deterministic pushdown automaton that accepts the language $\{wcw^R \mid w \in \{a,b\}^*\}$.

5. Show that the language $\{a^n b^m \mid n \neq m\}$ is deterministic context-free.

6. Show that the language $L = \{w \in \{a,b\}^* \mid n_a(w) > n_b(w)\}$ is deterministic context-free.

7. Let $M = (Q, \Sigma, \Lambda, q_0, \partial, F)$ be a pushdown automaton. Define $L'(M)$ to be the language $L'(M) = \{w \in \Sigma^* \mid$ it is possible for M to start in state q_0, read all of w, and end in an accepting state$\}$. $L'(M)$ differs from $L(M)$ in that for $w \in L'(M)$, we do not require that the stack be empty at the end of the computation.

a) Show that there is a pushdown automaton M' such that $L(M') = L'(M)$.

b) Show that a language L is context-free if and only if there is a pushdown automaton M such that $L = L'(M)$.

c) Identify the language $L'(M)$ for each of the automata in Exercise 1.

8. Let L be a regular language over an alphabet Σ, and let K be a context-free language over the same alphabet. Let $M = (Q, \Sigma, q_0, \delta, F)$ be a DFA that accepts L, and let $N = (P, \Sigma, \Lambda, p_0, \partial, E))$ be a pushdown automaton that accepts K. Show that the language $L \cap K$ is context-free by constructing a pushdown automaton that accepts $L \cap K$. The pushdown automaton can be constructed as a "cross product" of M and N in which the set of states is $Q \times P$. The construction is analogous to the proof that the intersection of two regular languages is regular, as outlined in Exercise 3.6.7.

4.5 Non-context-free Languages

We have seen that there are context-free languages that are not regular. The natural question arises, are there languages that are not context-free? It's easy to answer this question in the abstract: For a given alphabet Σ, there are uncountably many languages over Σ, but there are only countably many context-free languages over Σ. It follows that most languages are not context-free. However, this answer is not very satisfying since it doesn't give us any example of a specific language that is not context-free.

As in the case of regular languages, one way to show that a given language L is not context-free is to find some property that is shared by all context-free languages and then to show that L does not have that property. For regular languages, the Pumping Lemma gave us such a property. It turns out that there is a similar Pumping Lemma for context-free languages. The proof of this lemma uses parse trees. In the proof, we will need a way of representing abstract parse trees, without showing all the details of the tree. The picture

represents a parse tree which has the non-terminal symbol A at its root and the string x along the "bottom" of the tree. (That is, x is the string made up of all the symbols at the endpoints of the tree's branches, read from left to right.) Note that this could be a partial parse tree—something that could be a part of a larger tree. That is, we do not require A to be the start symbol of the grammar and we allow x to contain both terminal and non-terminal symbols. The string x, which is along the bottom of the tree, is referred to as the ***yield*** of the parse tree. Sometimes, we need to show more explicit detail in the tree. For example, the picture

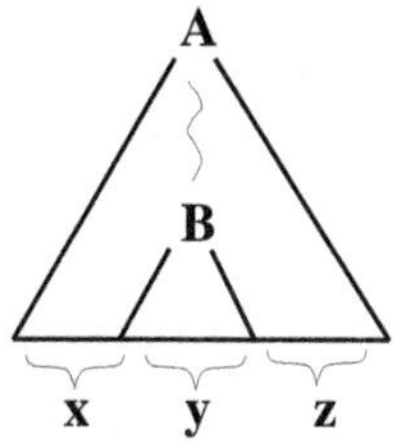

represents a parse tree in which the yield is the string xyz. The string y is the yield of a smaller tree, with root B, which is contained within the larger tree. Note that any of the strings x, y, or z could be the empty string.

We will also need the concept of the ***height*** of a parse tree. The height of a parse tree is the length of the longest path from the root of the tree to the tip of one of its branches.

Like the version for regular languages, the Pumping Lemma for context-free languages shows that any sufficiently long string in a context-free language contains a pattern that can be repeated to produce new strings that are also in the language. However, the pattern in this case is more complicated. For regular languages, the pattern arises because any sufficiently long path through a given DFA must contain a loop. For context-free languages, the pattern arises because in a sufficiently large parse tree, along a path from the root of the tree to the tip of one of its branches, there must be some non-terminal symbol that occurs more than once.

Theorem 4.7 (Pumping Lemma for Context-free Languages). *Suppose that L is a context-free language. Then there is an integer K such that any string $w \in L(G)$ with $|w| \geq K$ has the property that w can be written in the form $w = uxyzv$ where*

- *x and z are not both equal to the empty string;*
- *$|xyz| < K$; and*
- *For any $n \in \mathbb{N}$, the string ux^nyz^nv is in L.*

Proof. Let $G = (V, \Sigma, P, S)$ be a context-free grammar for the language L. Let N be the number of non-terminal symbols in G, plus 1. That is,

$N = |V| + 1$. Consider all possible parse trees for the grammar G with height less than or equal to N. (Include parse trees with any non-terminal symbol as root, not just parse trees with root S.) There are only finitely many such parse trees, and therefore there are only finitely many different strings that are the yields of such parse trees. Let K be an integer which is greater than the length of any such string.

Now suppose that w is any string in L whose length is greater than or equal to K. Then *any* parse tree for w must have height greater than N. (This follows since $|w| \geq K$ and the yield of any parse tree of height $\leq N$ has length less than K.) Consider a parse tree for w of minimal size, that is one that contains the smallest possible number of nodes. Since the height of this parse tree is greater than N, there is at least one path from the root of the tree to tip of a branch of the tree that has length greater than N. Consider the longest such path. The symbol at the tip of this path is a terminal symbol, but all the other symbols on the path are non-terminal symbols. There are at least N such non-terminal symbols on the path. Since the number of different non-terminal symbols is $|V|$ and since $N = |V| + 1$, some non-terminal symbol must occur twice on the path. In fact, some non-terminal symbol must occur twice among the bottommost N non-terminal symbols on the path. Call this symbol A. Then we see that the parse tree for w has the form shown here:

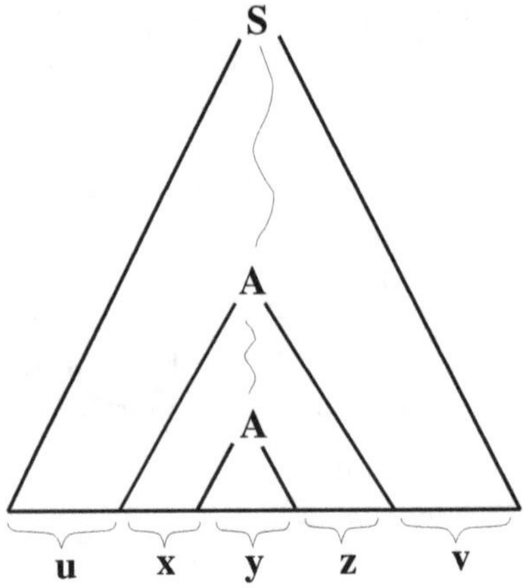

The structure of this tree breaks the string w into five substrings, as shown in the above diagram. We then have $w = uxyzv$. It only remains to show that x, y, and z satisfy the three requirements stated in the theorem.

Let T refer to the entire parse tree, let T_1 refer to the parse tree whose root is the upper A in the diagram, and let T_2 be the parse tree whose root is the lower A in the diagram. Note that the height of T_1 is less than or equal to N. (This follows from two facts: The path shown in T_1 from its root to its base has length less than or equal to N, because we chose the two occurrences of A to be among the N bottommost non-terminal symbols along the path in T from its root to its base. We know that there is no longer path from the root of T_1 to its base, since we chose the path in T to be the longest possible path from the root of T to its base.) Since any parse tree with height less than or equal to N has yield of length less than K, we see that $|xyz| < K$.

If we remove T_1 from T and replace it with a copy of T_2, the result is a parse tree with yield uyv, so we see that the string uyv is in the language L. Now, suppose that both x and z are equal to the empty string. In that case, $w = uyv$, so the tree we have created would be another parse tree for w. But this tree is smaller than T, so this would contradict the fact that T is the smallest parse tree for w. We see that x and z cannot both be the empty string.

If we remove T_2 from T and replace it with a copy of T_1, the result is a parse tree with yield ux^2yz^2v, so we see that $ux^2yz^2v \in L$. The two parse trees that we have created look like this:

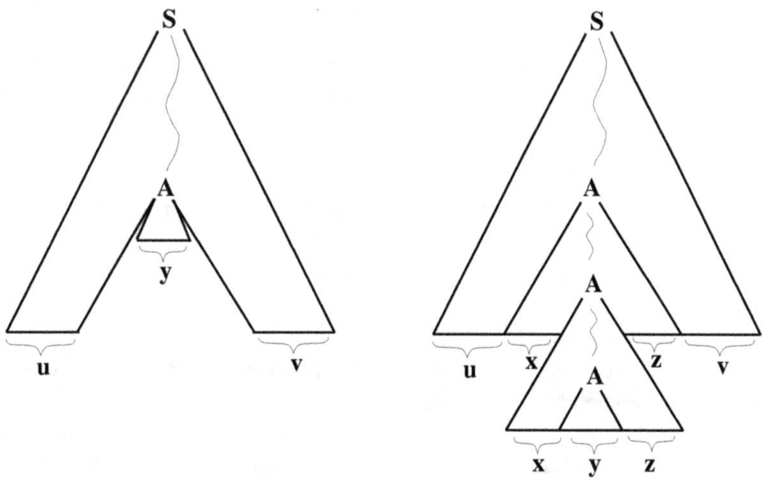

Furthermore, we can apply the process of replacing T_2 with a copy of T_1 to the tree on the right above to create a parse tree with yield ux^3yz^3v. Continuing in this way, we see that $ux^nyz^nv \in L$ for all $n \in \mathbb{N}$. This completes the proof of the theorem. □

Since this theorem guarantees that all context-free languages have a certain property, it can be used to show that specific languages are not context-free. The method is to show that the language in question does not have the property that is guaranteed by the theorem. We give two examples.

Corollary 4.8. *Let L be the language $\{a^n b^n c^n \mid n \in \mathbb{N}\}$. Then L is not a context-free language.*

Proof. We give a proof by contradiction. Suppose that L is context-free. Then, by the Pumping Lemma for Context-free Languages, there is an integer K such that every string $w \in L$ with $|w| \geq K$ can be written in the form $w = uxyzv$ where x and z are not both empty, $|xyz| < K$, and $ux^n yz^n v \in L$ for every $n \in \mathbb{N}$.

Consider the string $w = a^K b^K c^K$, which is in L, and write $w = uxyzv$, where u, x, y, z, and v satisfy the stated conditions. Since $|xyz| < K$, we see that if xyz contains an a, then it cannot contain a c. And if it contains a c, then it cannot contain an a. It is also possible that xyz is made up entirely of b's. In any of these cases, the string ux^2yz^2v cannot be in L, since it does not contain equal numbers of a's, b's, and c's. But this contradicts the fact that $ux^n yz^n v \in L$ for all $n \in \mathbb{N}$. This contradiction shows that the assumption that L is context-free is incorrect. □

Corollary 4.9. *Let Σ be any alphabet that contains at least two symbols. Let L be the language over Σ defined by $L = \{ss \mid s \in \Sigma^*\}$. Then L is not context-free.*

Proof. Suppose, for the sake of contradiction, that L is context-free. Then, by the Pumping Lemma for Context-free Languages, there is an integer K such that every string $w \in L$ with $|w| \geq K$ can be written in the form $w = uxyzv$ where x and z are not both empty, $|xyz| < K$, and $ux^n yz^n v \in L$ for every $n \in \mathbb{N}$.

Let a and b represent distinct symbols in Σ. Let $s = a^K ba^K b$ and let $w = ss = a^K ba^K ba^K ba^K b$, which is in L. Write $w = uxyzv$, where u, x, y, z, and v satisfy the stated conditions.

Since $|xyz| < K$, x and z can, together, contain no more than one b. If either x or y contains a b, then ux^2yz^2v contains exactly five b's. But any string in L is of the form rr for some string r and so contains an even number of b's. The fact that ux^2yz^2z contains five b's contradicts the fact that $ux^2yz^2v \in L$. So, we get a contradiction in the case where x or y contains a b.

Now, consider the case where x and y consist entirely of a's. Again since $|xyz| < K$, we must have either that x and y are both contained in the same group of a's in the string $a^K ba^K ba^K ba^K b$, or that x is contained in one group of a's and y is contained in the next. In either case, it is easy to check that the string ux^2yz^2v is no longer of the form rr for any string r, which contradicts the fact that $ux^2yz^2v \in L$.

Since we are led to a contradiction in every case, we see that the assumption that L is context-free must be incorrect. □

Now that we have some examples of languages that are not context-free, we can settle some other questions about context-free languages. In particular, we can show that the intersection of two context-free languages is not necessarily context-free and that the complement of a context-free language is not necessarily context-free.

Theorem 4.10. *The intersection of two context-free languages is not necessarily a context-free language.*

Proof. To prove this, it is only necessary to produce an example of two context-free languages L and M such that $L \cap M$ is not a context-free languages. Consider the following languages, defined over the alphabet $\Sigma = \{a, b, c\}$:

$$L = \{a^n b^n c^m \mid n \in \mathbb{N} \text{ and } m \in \mathbb{N}\}$$
$$M = \{a^n b^m c^m \mid n \in \mathbb{N} \text{ and } m \in \mathbb{N}\}$$

Note that strings in L have equal numbers of a's and b's while strings in M have equal numbers of b's and c's. It follows that strings in $L \cap M$ have equal numbers of a's, b's, and c's. That is,

$$L \cap M = \{a^n b^n c^n \mid n \in \mathbb{N}\}$$

We know from the above theorem that $L \cap M$ is not context-free. However, both L and M are context-free. The language L is generated by the context-free grammar

$$S \longrightarrow TC$$
$$C \longrightarrow cC$$
$$C \longrightarrow \varepsilon$$
$$T \longrightarrow aTb$$
$$T \longrightarrow \varepsilon$$

and M is generated by a similar context-free grammar. □

Corollary 4.11. *The complement of a context-free language is not necessarily context-free.*

Proof. Suppose for the sake of contradiction that the complement of every context-free language is context-free.

Let L and M be two context-free languages over the alphabet Σ. By our assumption, the complements $\overline{L}$ and $\overline{M}$ are context-free. By Theorem 4.3, it follows that $\overline{L} \cup \overline{M}$ is context-free. Applying our assumption once again, we have that $\overline{\overline{L} \cup \overline{M}}$ is context-free. But $\overline{\overline{L} \cup \overline{M}} = L \cap M$, so we have that $L \cap M$ is context-free.

We have shown, based on our assumption that the complement of any context-free language is context-free, that the intersection of any two context-free languages is context-free. But this contradicts the previous theorem, so we see that the assumption cannot be true. This proves the theorem. □

Note that the preceding theorem and corollary say only that $L \cap M$ is not context-free for *some* context-free languages L and M and that $\overline{L}$ is not context-free for *some* context-free language L. There are, of course, many examples of context-free languages L and M for which $L \cap M$ and $\overline{L}$ *are* in fact context-free.

Even though the intersection of two context-free languages is not necessarily context-free, it happens that the intersection of a context-free language with a *regular* language is always context-free. This is not difficult to show, and a proof is outlined in Exercise 4.4.8. I state it here without proof:

Theorem 4.12. *Suppose that L is a context-free language and that M is a regular language. Then $L \cap M$ is a context-free language.*

For example, let L and M be the languages defined by $L = \{w \in \{a,b\}^* \mid w = w^R\}$ and $M = \{w \in \{a,b\}^* \mid$ the length of w is a multiple of 5$\}$. Since L is context-free and M is regular, we know that $L \cap M$ is context-free. The language $L \cap M$ consists of every palindrome over the alphabet $\{a,b\}$ whose length is a multiple of five.

This theorem can also be used to show that certain languages are *not* context-free. For example, consider the language $L = \{w \in \{a,b,c\}^* \mid n_a(w) = n_b(w) = n_c(w)\}$. (Recall that $n_x(w)$ is the number of times that the symbol x occurs in the string w.) We can use a proof by contradiction to show that L is not context-free. Let M be the regular language defined by the regular expression $a^*b^*c^*$. It is clear that $L \cap M = \{a^nb^nc^n \mid n \in \mathbb{N}\}$. If L were context-free, then, by the previous theorem, $L \cap M$ would be context-free. However, we know from Theorem 4.8 that $L \cap M$ is not context-free. So we can conclude that L is not context-free.

Exercises

1. Show that the following languages are not context-free:
   a) $\{a^n b^m c^k \mid n > m > k\}$
   b) $\{w \in \{a,b,c\}^* \mid n_a(w) > n_b(w) > n_c(w)\}$
   c) $\{www \mid w \in \{a,b\}^*\}$
   d) $\{a^n b^m c^k \mid n, m \in \mathbb{N} \text{ and } k = m * n\}$
   e) $\{a^n b^m \mid m = n^2\}$

2. Show that the languages $\{a^n \mid \text{n is a prime number}\}$ and $\{a^{n^2} \mid n \in \mathbb{N}\}$ are not context-free. (In fact, it can be shown that a language over the alphabet $\{a\}$ is context-free if and only if it is regular.)

3. Show that the language $\{w \in \{a,b\}^* \mid n_a(w) = n_b(w)$ and w contains the string *baaab* as a substring$\}$ is context-free.

4. Suppose that M is any finite language and that L is any context-free language. Show that the language $L \smallsetminus M$ is context-free. (Hint: Any finite language is a regular language.)

4.6 General Grammars

At the beginning of this chapter the general idea of a grammar as a set of rewriting or production rules was introduced. For most of the chapter, however, we have restricted our attention to context-free grammars, in which production rules must be of the form $A \longrightarrow x$ where A is a non-terminal symbol. In this section, we will consider general grammars, that is, grammars in which there is no such restriction on the form of production rules. For a general grammar, a production rule has the form $u \longrightarrow x$, where u is string that can contain both terminal and non-terminal symbols. For convenience, we will assume that u contains at least one non-terminal symbol, although even this restriction could be lifted without changing the class of languages that can be generated by grammars. Note that a context-free grammar is, in fact, an example of a general grammar, since production rules in a general grammar are allowed to be of the form $A \longrightarrow x$. They just don't have to be of this form. I will use the unmodified term ***grammar*** to refer to general grammars.[2] The definition of grammar is identical to the definition of context-free grammar, except for the form of the production rules:

[2]There is another special type of grammar that is intermediate between context-free grammars and general grammars. In a so-called ***context-sensitive grammar***, every production rule is of the form $u \longrightarrow x$ where $|x| \geq |u|$. We will not cover context-sensitive grammars in this text.

Definition 4.6. A ***grammar*** is a 4-tuple (V, Σ, P, S), where:

1. V is a finite set of symbols. The elements of V are the non-terminal symbols of the grammar.
2. Σ is a finite set of symbols such that $V \cap \Sigma = \emptyset$. The elements of Σ are the terminal symbols of the grammar.
3. P is a set of production rules. Each rule is of the form $u \longrightarrow x$ where u and x are strings in $(V \cup \Sigma)^*$ and u contains at least one symbol from V.
4. $S \in V$. S is the start symbol of the grammar.

Suppose G is a grammar. Just as in the context-free case, the language generated by G is denoted by $L(G)$ and is defined as $L(G) = \{x \in \Sigma^* \mid S \Longrightarrow_G^* x\}$. That is, a string x is in $L(G)$ if and only if x is a string of terminal symbols and there is a derivation that produces x from the start symbol, S, in one or more steps.

The natural question is whether there are languages that can be generated by general grammars but that cannot be generated by context-free languages. We can answer this question immediately by giving an example of such a language. Let L be the language $L = \{w \in \{a,b,c\}^* \mid n_a(w) = n_b(w) = n_c(w)\}$. We saw at the end of the last section that L is not context-free. However, L is generated by the following grammar:

$$\begin{aligned}
S &\longrightarrow SABC \\
S &\longrightarrow \varepsilon \\
AB &\longrightarrow BA \\
BA &\longrightarrow AB \\
AC &\longrightarrow CA \\
CA &\longrightarrow AC \\
BC &\longrightarrow CB \\
CB &\longrightarrow BC \\
A &\longrightarrow a \\
B &\longrightarrow b \\
C &\longrightarrow c
\end{aligned}$$

For this grammar, the set of non-terminals is $\{S, A, B, C\}$ and the set of terminal symbols is $\{a, b, c\}$. Since both terminals and non-terminal symbols can occur on the left-hand side of a production rule in a general grammar, it is not possible, in general, to determine which symbols are non-terminal and which are terminal just by looking at the list of production rules. However, I will follow the convention that uppercase letters are always non-terminal symbols. With this convention, I can continue to specify a grammar simply by listing production rules.

The first two rules in the above grammar make it possible to produce the strings ε, ABC, $ABCABC$, $ABCABCABC$, and so on. Each of these strings contains equal numbers of A's, B's, and C's. The next six rules allow the order of the non-terminal symbols in the string to be changed. They make it possible to arrange the A's, B's, and C's into any arbitrary order. Note that these rules could not occur in a context-free grammar. The last three rules convert the non-terminal symbols A, B, and C into the corresponding terminal symbols a, b, and c. Remember that all the non-terminals must be eliminated in order to produce a string in $L(G)$. Here, for example, is a derivation of the string *baabcc* using this grammar. In each line, the string that will be replaced on the next line is underlined.

$$
\begin{aligned}
S &\Longrightarrow \underline{S}ABC \\
&\Longrightarrow \underline{S}ABCABC \\
&\Longrightarrow \underline{AB}CABC \\
&\Longrightarrow BA\underline{CA}BC \\
&\Longrightarrow BAA\underline{CB}C \\
&\Longrightarrow \underline{B}AABCC \\
&\Longrightarrow b\underline{A}ABCC \\
&\Longrightarrow ba\underline{A}BCC \\
&\Longrightarrow baa\underline{B}CC \\
&\Longrightarrow baab\underline{C}C \\
&\Longrightarrow baabc\underline{C} \\
&\Longrightarrow baabcc
\end{aligned}
$$

We could produce any string in L in a similar way. Of course, this only shows that $L \subseteq L(G)$. To show that $L(G) \subseteq L$, we can observe that for any string w such that $S \Longrightarrow^* w$, $n_A(w) + n_a(w) = n_B(w) + n_b(w) = n_C(w) + n_c(w)$. This follows since the rule $S \Longrightarrow SABC$ produces strings in which $n_A(w) = n_B(w) = n_C(w)$, and no other rule changes any of the quantities $n_A(w)+n_a(w)$, $n_B(w)+n_b(w)$, or $n_C(w)+n_c(w)$. After applying these rules to produce a string $x \in L(G)$, we must have that $n_A(x)$, $n_B(x)$, and $n_C(x)$ are zero. The fact that $n_a(x) = n_b(x) = n_c(x)$ then follows from the fact that $n_A(x) + n_a(x) = n_B(x) + n_b(x) = n_C(x) + n_c(x)$. That is, $x \in L$.

Our first example of a non-context-free language was $\{a^n b^n c^n \mid n \in \mathbb{N}\}$. This language can be generated by a general grammar similar to the previous example. However, it requires some cleverness to force the a's, b's, and c's into the correct order. To do this, instead of allowing A's, B's, and C's to transform themselves spontaneously into a's, b's, and c's, we use additional non-terminal symbols to transform them only after they are in the correct position. Here is a grammar that does this:

$$\begin{aligned}
S &\longrightarrow SABC \\
S &\longrightarrow X \\
BA &\longrightarrow AB \\
CA &\longrightarrow AC \\
CB &\longrightarrow BC \\
XA &\longrightarrow aX \\
X &\longrightarrow Y \\
YB &\longrightarrow bY \\
Y &\longrightarrow Z \\
ZC &\longrightarrow cZ \\
Z &\longrightarrow \varepsilon
\end{aligned}$$

Here, the first two rules produce one of the strings X, $XABC$, $XABCABC$, $XABCABCABC$, and so on. The next three rules allow A's to move to the left and C's to move to the right, producing a string of the form $XA^nB^nC^n$, for some $n \in \mathbb{N}$. The rule $XA \longrightarrow aX$ allows the X to move through the A's from left to right, converting A's to a's as it goes. After converting the A's, the X can be transformed into a Y. The Y will then move through the B's, converting them to b's. Then, the Y is transformed into a Z, which is responsible for converting C's to c's. Finally, an application of the rule $Z \longrightarrow \varepsilon$ removes the Z, leaving the string $a^nb^nc^n$.

Note that if the rule $X \longrightarrow Y$ is applied before all the A's have been converted to a's, then there is no way for the remaining A's to be converted to a's or otherwise removed from the string. This means that the derivation has entered a dead end, which can never produce a string that consists of terminal symbols only. The only derivations that can produce strings in the language generated by the grammar are derivations in which the X moves past all the A's, converting them all to a's. At this point in the derivation, the string is of the form a^nXu where u is a string consisting entirely of B's and C's. At this point, the rule $X \longrightarrow Y$ can be applied, producing the string a^nYu. Then, if a string of terminal symbols is ever to be produced, the Y must move past all the B's, producing the string $a^nb^nYC^n$. You can see that the use of three separate non-terminals, X, Y, and Z, is essential for forcing the symbols in $a^nb^nc^n$ into the correct order.

For one more example, consider the language $\{a^{n^2} \mid n \in \mathbb{N}\}$. Like the other languages we have considered in this section, this language is not context-free. However, it can be generated by a grammar. Consider the grammar

$$\begin{aligned} S &\longrightarrow DTE \\ T &\longrightarrow BTA \\ T &\longrightarrow \varepsilon \\ BA &\longrightarrow AaB \\ Ba &\longrightarrow aB \\ BE &\longrightarrow E \\ DA &\longrightarrow D \\ Da &\longrightarrow aD \\ DE &\longrightarrow \varepsilon \end{aligned}$$

The first three rules produce all strings of the form DB^nA^nE, for $n \in \mathbb{N}$. Let's consider what happens to the string DB^nA^nE as the remaining rules are applied. The next two rules allow a B to move to the right until it reaches the E. Each time the B passes an A, a new a is generated, but a B will simply move past an a without generating any other characters. Once the B reaches the E, the rule $BE \longrightarrow E$ makes the B disappear. Each B from the string DB^nA^nE moves past n A's and generates n a's. Since there are n B's, a total of n^2 a's are generated. Now, the only way to get rid of the D at the beginning of the string is for it to move right through all the A's and a's until it reaches the E at the end of the string. As it does this, the rule $DA \longrightarrow D$ eliminates all the A's from the string, leaving the string $a^{n^2}DE$. Applying the rule $DE \longrightarrow \varepsilon$ to this gives a^{n^2}. This string contains no non-terminal symbols and so is in the language generated by the grammar. We see that every string of the form a^{n^2} is generated by the above grammar. Furthermore, only strings of this form can be generated by the grammar.

Given a fixed alphabet Σ, there are only countably many different languages over Σ that can be generated by grammars. Since there are uncountably many different languages over Σ, we know that there are many languages that cannot be generated by grammars. However, it is surprisingly difficult to find an actual example of such a language.

As a first guess, you might suspect that just as $\{a^nb^n \mid n \in \mathbb{N}\}$ is an example of a language that is not regular and $\{a^nb^nc^n \mid n \in \mathbb{N}\}$ is an example of a language that is not context-free, so $\{a^nb^nc^nd^n \mid n \in \mathbb{N}\}$ might be an example of a language that cannot be generated by any grammar. However, this is not the case. The same technique that was used to produce a grammar that generates $\{a^n b^n c^n \mid n \in \mathbb{N}\}$ can also be used to produce a grammar for $\{a^n b^n c^n d^n \mid n \in \mathbb{N}\}$. In fact, the technique extends to similar languages based on any number of symbols.

Or you might guess that there is no grammar for the language $\{a^n \mid n$ is a prime number $\}$. Certainly, producing prime numbers doesn't seem like the kind of thing that we would ordinarily do with a grammar. Nevertheless, there is a grammar that generates this language. We will not actually write down the grammar, but we will eventually have a way to prove that it exists.

The language $\{a^{n^2} \mid n \in \mathbb{N}\}$ really doesn't seem all that "grammatical" either, but we produced a grammar for it above. If you think about how this grammar works, you might get the feeling that its operation is more like "computation" than "grammar." This is our clue. A grammar can be thought of as a kind of program, albeit one that is executed in a nondeterministic fashion. It turns out that general grammars are precisely as powerful as any other general-purpose programming language, such as Java or C++. More exactly, a language can be generated by a grammar if and only if there is a computer program whose output consists of a list containing all the strings and only the strings in that language. Languages that have this property are said to be ***recursively enumerable languages***. (This term as used here is *not* closely related to the idea of a recursive subroutine.) The languages that can be generated by general grammars are precisely the recursively enumerable languages. We will return to this topic in the next chapter.

It turns out that there are many forms of computation that are precisely equivalent in power to grammars and to computer programs, and no one has ever found any form of computation that is more powerful. This is one of the great discoveries of the twentieth century, and we will investigate it further in the next chapter.

Exercises

1. Find a derivation for the string *caabcb*, according to the first example grammar in this section. Find a derivation for the string *aabbcc*, according to the second example grammar in this section. Find a derivation for the string *aaaa*, according to the third example grammar in this section.
2. Consider the third sample grammar from this section, which generates the language $\{a^{n^2} \mid n \in \mathbb{N}\}$. Is the non-terminal symbol D necessary in this grammar? What if the first rule of the grammar were replaced by $S \longrightarrow TE$ and the last three rules were replaced by $A \longrightarrow \varepsilon$ and $E \longrightarrow \varepsilon$? Would the resulting grammar still generate the same language? Why or why not?

3. Find a grammar that generates the language $L = \{w \in \{a,b,c,d\}^* \mid n_a(w) = n_b(w) = n_c(w) = n_d(w)\}$. Let Σ be any alphabet. Argue that the language $\{w \in \Sigma^* \mid$ all symbols in Σ occur equally often in $w\}$ can be generated by a grammar.

4. For each of the following languages, find a grammar that generates the language. In each case, explain how your grammar works.

a) $\{a^n b^n c^n d^n \mid n \in \mathbb{N}\}$

b) $\{a^n b^m c^{nm} \mid n \in \mathbb{N} \text{ and } m \in \mathbb{N}\}$

c) $\{ww \mid w \in \{a,b\}^*\}$

d) $\{www \mid w \in \{a,b\}^*\}$

e) $\{a^{2^n} \mid n \in \mathbb{N}\}$

f) $\{w \in \{a,b,c\}^* \mid n_a(w) > n_b(w) > n_c(w)\}$

Chapter 5

Turing Machines and Computability

We saw hints at the end of the previous chapter that "computation" is a more general concept than we might have thought. General grammars, which at first encounter don't seem to have much to do with algorithms or computing, turn out to be able to do things that are similar to the tasks carried out by computer programs. In this chapter, we will see that general grammars are precisely equivalent to computer programs in terms of their computational power, and that both are equivalent to a particularly simple model of computation known as a ***Turing machine***. We shall also see that there are limits to what can be done by computing.

5.1 Turing Machines

Historically, the theoretical study of computing began before computers existed. One of the early models of computation was developed in the 1930s by the British mathematician, Alan Turing, who was interested in studying the theoretical abilities and limitations of computation. His model for computation is a very simple abstract computing machine which has come to be known as a ***Turing machine***. While Turing machines are not applicable in the same way that regular expressions, finite-state automata, and grammars are applicable, their use as a fundamental model for computation means that every computer scientist should be familiar with them, at least in a general way.

A Turing machine is really not much more complicated than a finite-state automaton or a pushdown automaton.[1] Like a FSA, a Turing machine has a finite number of possible states, and it changes from state to state as it computes. However, a Turing machine also has an infinitely long ***tape*** that it can use for input and output. The tape extends to infinity in both directions. The tape is divided into ***cells***, which are in one-to-one correspondence with the integers, $\mathbb{Z}$. Each cell can either be blank or it can hold a symbol from a specified alphabet. The Turing machine can move back and forth along this tape, reading and writing symbols and changing state. It can read only one cell at a time, and possibly write a new value in that cell. After doing this, it can change state and it can move by one cell either to the left or to the right. This is how the Turing machine computes. To use a Turing machine, you would write some input on its tape, start the machine, and let it compute until it halts. Whatever is written on the tape at that time is the output of the computation.

Although the tape is infinite, only a finite number of cells can be non-blank at any given time. If you don't like the idea of an infinite tape, you can think of a finite tape that can be extended to an arbitrarily large size as the Turing machine computes: If the Turing machine gets to either end of the tape, it will pause and wait politely until you add a new section of tape. In other words, it's not important that the Turing machine have an infinite amount of memory, only that it can use as much memory as it needs for a given computation, up to any arbitrarily large size. In this way, a Turing machine is like a computer that can ask you to buy it a new disk drive whenever it needs more storage space to continue a computation.[2]

A given Turing machine has a fixed, finite set of states. One of these states is designated as the ***start state***. This is the state in which the Turing machine begins a computation. Another special state is the ***halt state***. The Turing machine's computation ends when it enters its halt state. It is possible that a computation might never end because the machine never enters the halt state. This is analogous to an infinite loop in a computer program.

---

[1] In fact, Turing machines can be shown to be equivalent in their computational power to pushdown automata with two independent stacks.

[2] The tape of a Turing machine can be used to store arbitrarily large amounts of information in a straightforward way. Although we can imagine using an arbitrary amount of memory with a computer, it's not so easy. Computers aren't set up to keep track of unlimited amounts of data. If you think about how it might be done, you probably won't come with anything better than an infinite tape. (The problem is that computers use integer-valued addresses to keep track of data locations. If a limit is put on the number of bits in an address, then only a fixed, finite amount of data can be addressed. If no limit is put on the number of bits in an address, then we are right back to the problem of storing an arbitrarily large piece of data—just to represent an address!)

At each step in its computation, the Turing machine reads the contents of the tape cell where it is located. Depending on its state and the symbol that it reads, the machine writes a symbol (possibly the same symbol) to the cell, moves one cell either to the left or to the right, and (possibly) changes its state. The output symbol, direction of motion, and new state are determined by the current state and the input symbol. Note that either the input symbol, the output symbol, or both, can be blank. A Turing machine has a fixed set of ***rules*** that tell it how to compute. Each rule specifies the output symbol, direction of motion, and new state for some combination of current state and input symbol. The machine has a rule for every possible combination of current state and input symbol, except that there are no rules for what happens if the current state is the halt state. Of course, once the machine enters the halt state, its computation is complete and the machine simply stops.

I will use the character # to represent a blank in a way that makes it visible. I will always use h to represent the halt state. I will indicate the directions, left and right, with L and R, so that $\{L, R\}$ is the set of possible directions of motion. With these conventions, we can give the formal definition of a Turing machine as follows:

Definition 5.1. A ***Turing machine*** is a 4-tuple $(Q, \Lambda, q_0, \delta)$, where:

1. Q is a finite set of states, including the halt state, h.
2. Λ is an alphabet which includes the blank symbol, #.
3. $q_0 \in Q$ is the start state.
4. $\delta\colon (Q \smallsetminus \{h\}) \times \Lambda \to \Lambda \times \{L, R\} \times Q$ is the transition function. The fact that $\delta(q, \sigma) = (\tau, d, r)$ means that when the Turing machine is in state q and reads the symbol σ, it writes the symbol τ, moves one cell in the direction d, and enters state r.

Even though this is the formal definition, it's easier to work with a transition diagram representation of Turing machines. The transition diagram for a Turing machine is similar to the transition diagram for a DFA. However, there are no "accepting" states (only a halt state). Furthermore, there must be a way to specify the output symbol and the direction of motion for each step of the computation. We do this by labeling arrows with notations of the form (σ, τ, L) and (σ, τ, R), where σ and τ are symbols in the Turing machine's alphabet. For example,

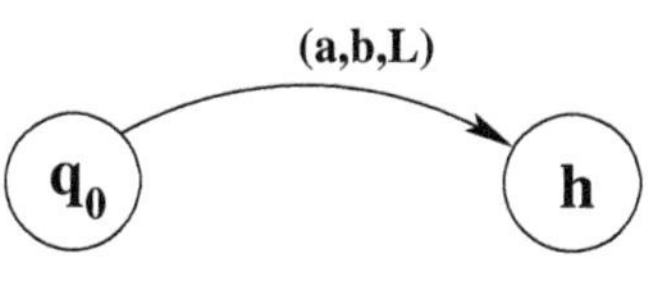

indicates that when the machine is in state q_0 and reads an a, it writes a b, moves left, and enters state h.

Here, for example, is a transition diagram for a simple Turing machine that moves to the right, changing a's to b's and *vice versa*, until it finds a c. It leaves blanks (#'s) unchanged. When and if the machine encounters a c, it moves to the left and halts:

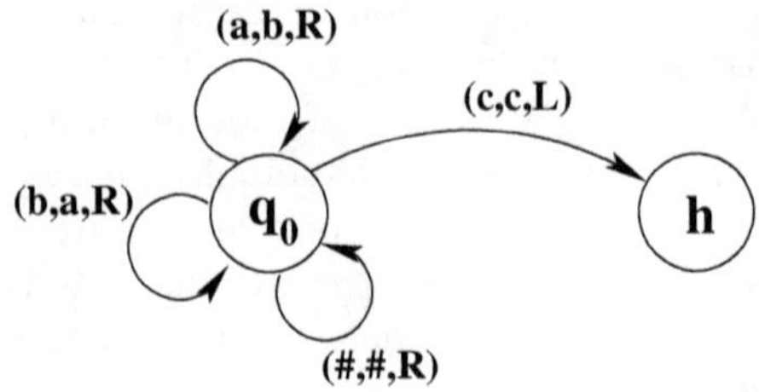

To simplify the diagrams, I will leave out any transitions that are not relevant to the computation that I want the machine to perform. You can assume that the action for any omitted transition is to write the same symbol that was read, move right, and halt.

For example, shown below is a transition diagram for a Turing machine that makes a copy of a string of a's and b's. To use this machine, you would write a string of a's and b's on its tape, place the machine on the first character of the string, and start the machine in its start state, q_0. When the machine halts, there will be two copies of the string on the tape, separated by a blank. The machine will be positioned on the first character of the leftmost copy of the string. Note that this machine uses c's and d's in addition to a's and b's. While it is copying the input string, it temporarily changes the a's and b's that it has copied to c's and d's, respectively. In this way it can keep track of which characters it has already copied. After the string has been copied, the machine changes the c's and d's back to a's and b's before halting.

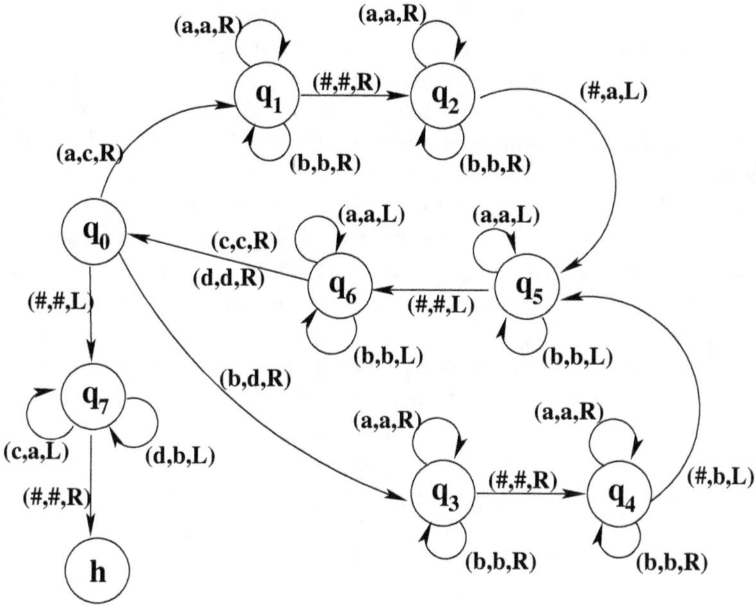

In this machine, state q_0 checks whether the next character is an a, a b, or a # (indicating the end of the string). States q_1 and q_2 add an a to the end of the new string, and states q_3 and q_4 do the same thing with a b. States q_5 and q_6 return the machine to the next character in the input string. When the end of the input string is reached, state q_7 will move the machine back to the start of the input string, changing c's and d's back to a's and b's as it goes. Finally, when the machine hits the # that precedes the input string, it moves to the right and halts. This leave it back at the first character of the input string. It would be a good idea to work through the execution of this machine for a few sample input strings. You should also check that it works even for an input string of length zero.

Our primary interest in Turing machines is as language processors. Suppose that w is a string over an alphabet Σ. We will assume that Σ does not contain the blank symbol. We can use w as input to a Turing machine $M = (Q, \Lambda, q_0, \delta)$ provided that $\Sigma \subseteq \Lambda$. To use w as input for M, we will write w on M's tape and assume that the remainder of the tape is blank. We place the machine on the cell containing the first character of the string, except that if $w = \varepsilon$ then we simply place the machine on a completely blank tape. Then we start the machine in its initial state, q_0, and see what computation it performs. We refer to this setup as "running M with input w."

When M is run with input w, it is possible that it will just keep running forever without halting. In that case, it doesn't make sense to ask about the output of the computation. Suppose however that M does halt on input w. Suppose, furthermore, that when M halts, its tape is blank except for a string x of non-blank symbols, and that the machine is located on the first character of x. In this case, we will say that "M halts with output x." In addition, if M halts with an entirely blank tape, we say that "M halts with output ε." Note that when we run M with input w, one of three things can happen: (1) M might halt with some string as output; (1) M might fail to halt; or (3) M might halt in some configuration that doesn't count as outputting any string.

The fact that a Turing machine can produce an output value allows us for the first time to deal with computation of *functions*. A function $f\colon A \to B$ takes an input value in the set A and produces an output value in the set B. If the sets are sets of strings, we can now ask whether the values of the function can be computed by a Turing machine. That is, is there a Turing machine M such that, given any string w in the domain of f as input, M will compute as its output the string $f(w)$. If this is that case, then we say that f is a Turing-computable function.

Definition 5.2. Suppose that Σ and Γ are alphabets that do not contain # and that f is a function from Σ^* to Γ^*. We say that f is ***Turing-computable*** if there is a Turing machine $M = (Q, \Lambda, q_0, \delta)$ such that $\Sigma \subseteq \Lambda$ and $\Gamma \subseteq \Lambda$ and for each string $w \in \Sigma^*$, when M is run with input w, it halts with output $f(w)$. In this case, we say that M **computes** the function f.

For example, let $\Sigma = \{a\}$ and define $f\colon \Sigma^* \to \Sigma^*$ by $f(a^n) = a^{2n}$, for $n \in \mathbb{N}$. Then f is Turing-computable since it is computed by this Turing machine:

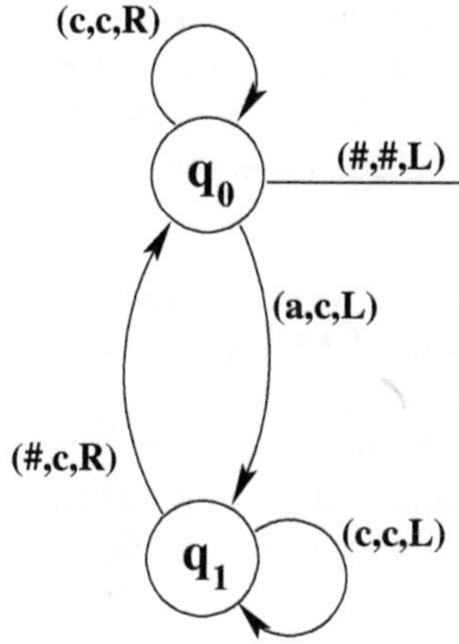

We can also use Turing machines to define "computable languages." There are actually two different notions of Turing-computability for languages. One is based on the idea of Turing-computability for functions. Suppose that Σ is an alphabet and that $L \subseteq \Sigma^*$. The ***characteristic function*** of L is the function $\chi_L \colon \Sigma^* \to \{0,1\}$ defined by the fact that $\chi_L(w) = 1$ if $w \in L$ and $\chi_L(w) = 0$ if $w \notin L$. Note that given the function χ_L, L can be obtained as the set $L = \{w \in \Sigma^* \mid \chi_L(w) = 1\}$. Given a language L, we can ask whether the corresponding function χ_L is Turing-computable. If so, then we can use a Turing machine to decide whether or not a given string w is in L. Just run the machine with input w. It will halt with output $\chi_L(w)$. (That is, it will halt and when it does so, the tape will be blank except for a 0 or a 1, and the machine will be positioned on the 0 or 1.) If the machine halts with output 1, then $w \in L$. If the machine halts with output 0, then $w \notin L$.

Definition 5.3. Let Σ be an alphabet that does not contain # and let L be a language over Σ. We say that L is ***Turing-decidable*** if there is a Turing machine $M = (Q, \Lambda, q_0, \delta)$ such that $\Sigma \subseteq \Lambda$, $\{0,1\} \subseteq \Lambda$, and for each $w \in \Sigma^*$, when M is run with input w, it halts with output $\chi_L(w)$. (That is, it halts with output 0 or 1, and the output is 0 if $w \notin L$ and is 1 if $w \in L$.) In this case, we say that M ***decides*** the language L.

The second notion of computability for languages is based on the interesting fact that it is possible for a Turing machine to run forever, without ever halting. Whenever we run a Turing machine M with input w, we can ask the question, will M ever halt or will it run forever? If M halts on input w, we will say that M "accepts" w. We can then look at all the strings over a given alphabet that are accepted by a given Turing machine. This leads to the notion of Turing-acceptable languages.

Definition 5.4. Let Σ be an alphabet that does not contain #, and let L be a language over Σ. We say that L is ***Turing-acceptable*** if there is a Turing machine $M = (Q, \Lambda, q_0, \delta)$ such that $\Sigma \subseteq \Lambda$, and for each $w \in \Sigma^*$, M halts on input w if and only if $w \in L$. In this case, we say that M ***accepts*** the language L.

It should be clear that any Turing-decidable language is Turing-acceptable. In fact, if L is a language over an alphabet Σ, and if M is a Turing machine that decides L, then it is easy to modify M to produce a Turing machine that accepts L. At the point where M enters the halt state with output 0, the new machine should enter a new state in which it simply moves to the right forever, without ever halting. Given an input $w \in \Sigma^*$, the modified machine will halt if and only if M halts with output 1, that is, if and only if $w \in L$.

Exercises

1. Let $\Sigma = \{a\}$. Draw a transition diagram for a Turing machine that computes the function $f: \Sigma^* \to \Sigma^*$ where $f(a^n) = a^{3n}$, for $n \in \mathbb{N}$. Draw a transition diagram for a Turing machine that computes the function $f: \Sigma^* \to \Sigma^*$ where $f(a^n) = a^{3n+1}$, for $n \in \mathbb{N}$.

2. Let $\Sigma = \{a, b\}$. Draw a transition diagram for a Turing machine that computes the function $f: \Sigma^* \to \Sigma^*$ where $f(w) = w^R$.

3. Suppose that Σ, Γ, and Ξ are alphabets and that $f: \Sigma^* \to \Gamma^*$ and $g: \Gamma^* \to \Xi^*$ are Turing-computable functions. Show that $g \circ f$ is Turing-computable.

4. We have defined computability for functions $f: \Sigma^* \to \Gamma^*$, where Σ and Γ are alphabets. How could Turing machines be used to define computable functions from $\mathbb{N}$ to $\mathbb{N}$? (Hint: Consider the alphabet $\Sigma = \{a\}$.)

5. Let Σ be an alphabet and let L be a language over Σ. Show that L is Turing-decidable if and only if its complement, $\overline{L}$, is Turing-decidable.

6. Draw a transition diagram for a Turing machine which decides the language $\{a^n b^n \mid n \in \mathbb{N}\}$. (Hint: Change the a's and b's to \$'s in pairs.) Explain in general terms how to make a Turing machine that decides the language $\{a^n b^n c^n \mid n \in \mathbb{N}\}$.

7. Draw a transition diagram for a Turing machine which decides the language $\{a^n b^m \mid n > 0 \text{ and } m \text{ is a multiple of } n\}$. (Hint: Erase n b's at a time.)

8. Based on your answer to the previous problem and the copying machine presented in this section, describe in general terms how you would build a Turing machine to decide the language $\{a^p \mid p \text{ is a prime number}\}$.

9. Let $g: \{a\}^* \to \{0,1\}^*$ be the function such that for each $n \in \mathbb{N}$, $g(a^n)$ is the representation of n as a binary number. Draw a transition diagram for a Turing machine that computes g.

5.2 Computability

At this point, it would be useful to look at increasingly complex Turing machines, which compute increasingly complex functions and languages. Although Turing machines are very simple devices, it turns out that they can perform very sophisticated computations. In fact, any computation that can be carried out by a modern digital computer—even one with an unlimited amount of memory—can be carried out by a Turing machine. Although it is not something that can be proved, it is widely believed that anything that can reasonably be called "computation" can be done by a Turing machine. This claim is known as the ***Church-Turing Thesis***.

We do not have time to look at enough examples to convince you that Turing machines are as powerful as computers, but the proof reduces to the fact that computers are actually fairly simple in their basic operation. Everything that a computer does comes down to copying data from one place to another, making simple comparisons between two pieces of data, and performing some basic arithmetic operations. It's possible for Turing machines to do all these things. In fact, it's possible to build a Turing machine to simulate the step-by-step operation of a given computer. Doing so proves that the Turing machine can do any computation that the computer could do, although it will, of course, work much, much more slowly.

We can, however, look briefly at some other models of computation and see how they compare with Turing machines. For example, there are various ways in which we might try to increase the power of a Turing machine. For example, consider a ***two-tape Turing machine*** that has two tapes, with a read/write head on each tape. In each step of its computation, a two-tape Turing machine reads the symbols under its read/write heads on both tapes. Based on these symbols and on its current state, it can write a new symbol onto each tape, independently move the read/write head on each tape one cell to the left or right, and change state.

It might seem that with two tapes available, two-tape Turing machines might be able to do computations that are impossible for ordinary one-tape machines. In fact, though, this is not the case. The reason, again, is simulation: Given any two-tape Turing machine, it is possible to build a one-tape Turing machine that simulates the step-by-step computation of the two-tape machine. Let M be a two-tape Turing machine. To simulate M with a one-tape machine, K, we must store the contents of both of M's tapes on one tape, and we must keep track of the positions of both of M's read/write heads. Let `@` and `$` be symbols that are not in the alphabet of M. The `@` will be used to mark the position of a read/write head, and the `$` will be used to delimit the parts of K's tape that represent the two tapes of M. For example, suppose that one of M's tapes contains the symbols "`abb##cca`" with the read/write head on the first `b`, and that the other tape contains "`01#111#001`" with the read/write head on the final `1`. This configuration would be represented on K's tape as "`$a@bb##cca$01#111#00@1$`". To simulate one step of M's computation, K must scan its entire tape, looking for the `@`'s and noting the symbol to the right of each `@`. Based on this information, K can update its tape and its own state to reflect M's new configuration after one step of computation. Obviously, K will take more steps than M and it will operate much more slowly, but this argument makes it clear that one-tape Turing machines can do anything that can be done by two-tape machines.

We needn't stop there. We can imagine n-tape Turing machines, for $n > 2$. We might allow a Turing machine to have multiple read/write heads that move independently on each tape. We could even allow two or three-dimensional tapes. None of this makes any difference as far as computational power goes, since each type of Turing machine can simulate any of the other types.[3]

We have used Turing machines to define Turing-acceptable languages and Turing-decidable languages. The definitions seem to depend very much on the peculiarities of Turing machines. But the same classes of languages can be defined in other ways. For example, we could use programs running on an idealized computer, with an unlimited amount of memory, to accept or decide languages. Or we could use n-tape Turing machines. The resulting classes of languages would be exactly the same as the Turing-acceptable and Turing-decidable languages.

We could look at other ways of specifying languages "computationally." One of the most natural is to imagine a Turing machine or computer program that runs forever and outputs an infinite list of strings over some alphabet Σ. In the case of Turing machines, it's convenient to think of a two-tape Turing machine that lists the strings on its second tape. The strings in the list form a language over Σ. A language that can be listed in this way is said to be ***recursively enumerable***. Note that we make no assumption that the strings must be listed in any particular order, and we

---

[3]We can also define ***non-deterministic Turing machines*** that can have several possible actions at each step. Non-deterministic Turing machines cannot be used to compute functions, since a function can have only one possible output for any given input. However, they can be used to accept languages. We say that a non-deterministic Turing machine accepts a language L is it is *possible* for the machine to halt on input w if and only if $w \in L$. The class of languages accepted by non-deterministic Turing machines is the same as the class of languages accepted by deterministic Turing machines. So, non-determinism does not add any computational power.

allow the same string to appear in the output any number of times. Clearly, a recursively enumerable language is "computable" in some sense. Perhaps we have found a new type of computable language. But no—it turns out that we have just found another way of describing the Turing-acceptable languages. The following theorem makes this fact official and adds one more way of describing the same class of languages:

Theorem 5.1. *Let Σ be an alphabet and let L be a language over Σ. Then the following are equivalent:*

1. *There is a Turing machine that accepts L.*
2. *There is a two-tape Turing machine that runs forever, making a list of strings on its second tape, such that a string w is in the list if and only if $w \in L$.*
3. *There is a Turing-computable function $f\colon \{a\}^* \to \Sigma^*$ such that L is the range of the function f.*

While I will not give a complete, formal proof of this theorem, it's not too hard to see why it is true. Consider a language that satisfies property 3. We can use the fact that L is the range of a Turing-computable function, f, to build a two-tape Turing machine that lists L. The machine will consider each of the strings a^n, for $n \in \mathbb{N}$, in turn, and it will compute $f(a^n)$ for each n. Once the value of $f(a^n)$ has been computed, it can be copied onto the machine's second tape, and the machine can move on to do the same with a^{n+1}. This machine writes all the elements of L (the range of f) onto its second tape, so L satisfies property 2. Conversely, suppose that there is a two-tape Turing machine, M, that lists L. Define a function $g\colon \{a\}^* \to \Sigma^*$ such that for $n \in \mathbb{N}$, $g(a^n)$ is the $(n+1)^{th}$ item in the list produced by M. Then the range of g is L, and g is Turing-computable since it can be computed as follows: On input a^n, simulate the computation of M until it has produced $n+1$ strings, then halt, giving the $(n+1)^{th}$ string as output. This shows that property 2 implies property 3, so these properties are in fact equivalent.

We can also check that property 2 is equivalent to property 1. Suppose that L satisfies property 2. Consider a two-tape Turing machine, T, that lists the elements of L. We must build a Turing machine, M, which accepts L. We do this as follows: Given an input $w \in \Sigma^*$, M will simulate the computation of T. Every time the simulated T produces a string in the list, M compares that string to w. If they are the same, M halts. If $w \in L$, eventually it will be produced by T, so M will eventually halt. If $w \notin L$, then it will never turn up in the list produced by T, so M will never halt. Thus, M accepts the language L. This shows that property 2 implies property 1.

The fact that property 1 implies property 2 is somewhat harder to see. First, we note that it is possible for a Turing machine to generate every possible string in Σ^*, one-by-one, in some definite order (such as order of increasing length, with something like alphabetical order for strings of the same length). Now, suppose that L is Turing-acceptable and that M is a Turing machine that accepts L. We need a two-tape Turing machine, T that makes a list of all the elements of L. Unfortunately, the following idea does *not* work: Generate each of the elements in Σ^* in turn, and see whether M accepts it. If so, then add it to the list on the second tape. It looks like we have a machine that lists all the elements of L. The problem is that the only way for T to "see whether M accepts" a string is to simulate the computation of M. Unfortunately, as soon as we try this for any string w that is not in L, the computation never ends! T will get stuck in the simulation and will never even move on to the next string. To avoid this problem, T must simulate multiple computations of M at the same time. T can keep track of these computations in different regions of its first tape (separated by \$'s). Let the list of all strings in Σ^* be $x_1, x_2, x_3, \ldots$. Then T should operate as follows:

1. Set up the simulation of M on input x_1, and simulate one step of the computation for x_1
2. Set up the simulation of M on input x_2, and simulate one step of the computation for x_1 and one step of the computation for x_2.
3. Set up the simulation of M on input x_3, and simulate one step of each of the computations, for x_1, x_2, and x_3.

...

n. Set up the simulation of M on input x_n, and simulate one step of each of the computations, for $x_1, x_2, \ldots, x_n$.

and so on. Each time one of the computations halts, T should write the corresponding x_i onto its second tape. Over the course of time, T simulates the computation of M for each input $w \in \Sigma^*$ for an arbitrary number of steps. If $w \in L$, the simulated computation will eventually end and w will appear on T's second tape. On the other hand, if $w \notin L$, then the simulated computation will never end, so w will not appear in the list. So we see that T does in fact make a list of all the elements, and only the elements of L. This completes an outline of the proof of the theorem.

Next, we compare Turing machines to a completely different method of specifying languages: general grammars. Suppose $G = (V, \Sigma, P, S)$ is a general grammar and that L is the language generated by G. Then there is a Turing machine, M, that accepts the same language, L. The alphabet for M will be $V \cup \Sigma \cup \{\$,\#\}$, where \$ is a symbol that is not in $V \cup \Sigma$. (We also assume that # is not in $V \cup \Sigma$.) Suppose that M is started with input w, where $w \in \Sigma^*$. We have to design M so that it will halt if and only if $w \in L$. The idea is to have M find each string that can be derived from the start symbol S. The strings will be written to M's tape and separated by \$'s. M can begin by writing the start symbol, S, on its tape, separated from w by a \$. Then it repeats the following process indefinitely: For each string on the tape and for each production rule, $x \longrightarrow y$, of G, search the string for occurrences of x. When one is found, add a \$ to the end of the tape and copy the string to the end of the tape, replacing the occurrence of x by y. The new string represents the results of applying the production rule $x \longrightarrow y$ to the string. Each time M produces a new string, it compares that string to w. If they are equal, then M halts. If w is in fact in L, then eventually M will produce the string w and will halt. Conversely, if w is not in L, then M will go on producing strings forever without ever finding w, so M will never halt. This shows that, in fact, the language L is accepted by M.

Conversely, suppose that L is a language over an alphabet Σ, and that L is Turing-acceptable. Then it is possible to find a grammar G that generates L. To do this, it's convenient to use the fact that, as discussed above, there is a Turing-computable function $f\colon \{a\}^* \to \Sigma$ such that L is the range of f. Let $M = (Q, \Lambda, q_0, \delta)$ be a Turing machine that computes the function f. We can build a grammar, G, that imitates the computations performed by M. The idea is that most of the production rules of G will imitate steps in the computation of M. Some additional rules are added to get things started, to clean up, and to otherwise bridge the conceptual gap between grammars and Turing machines.

The terminal symbols of G will be the symbols from the alphabet, Σ. For the non-terminal symbols, we use: the states of M, every member of Λ that is not in Σ, two special symbols $<$ and $>$, and two additional symbols S and A. (We can assume that all these symbols are distinct.) S will be the start symbol of G. As for production rules, we begin with the following three rules:

$$S \longrightarrow <q_0A>$$
$$A \longrightarrow aA$$
$$A \longrightarrow \varepsilon$$

These rules make it possible to produce any string of the form $<q_0a^n>$. This is the only role that S and A play in the grammar. Once we've gotten rid of S and A, strings of the remaining terminal and non-terminal symbols represent configurations of the Turing machine M. The string will contain exactly one of the states of M (which is, remember, one of the non-terminal symbols of G). This tells us which state M is in. The position of the state-symbol tells us where M is positioned on the tape: the state-symbol is located in the string to the left of the symbol on which M is positioned. And the special symbols $<$ and $>$ just represent the beginning and the end of a portion of the tape of M. So, the initial string $<q_0a^n>$ represents a configuration in which M is in its start state, and is positioned on the first a in a string of n a's. This is the starting configuration of M when it is run with input a^n.

Now, we need some production rules that will allow the grammar to simulate the computations performed by M. For each state q_i and each symbol $\sigma \in \Lambda$, we need a production rule that imitates the transition rule $\delta(q_i, \sigma) = (\tau, d, q_j)$. If $d = R$, that is if the machine moves to the right, then all we need is the rule

$$q_i\sigma \longrightarrow \tau q_j$$

This represents that fact that M converts the σ to a τ, moves to the right, and changes to state q_j. If $d = L$, that is if the machine moves to the left, then we will need several rules—one rule for each $\lambda \in \Lambda$, namely

$$\lambda q_i\sigma \longrightarrow q_j\lambda\tau$$

This rule says that M changes the σ to a τ, moves left, and changes to state q_j. The λ doesn't affect the application of the rule, but is necessary to represent the fact that M moves left.

Each application of one of these rules represents one step in the computation of M. There is one remaining requirement for correctly simulating M. Since M's tape contains an infinite number of cells and we are only representing a finite portion of that tape, we need a way to add and remove #'s at the ends of the string. We can use the following four rules to do this:

$$\begin{aligned} < &\longrightarrow <\# \\ <\# &\longrightarrow < \\ > &\longrightarrow \#> \\ \#> &\longrightarrow > \end{aligned}$$

These rules allow blank symbols to appear at the ends of the string when they are needed to continue the computation, and to disappear from the ends of the string whenever we like.

Now, suppose that w is some element of L. Then $w = f(a^n)$ for some $n \in \mathbb{N}$. We know that on input a^n, M halts with output w. If we translate the computation of M into the corresponding sequence of production rules in G, we see that for the grammar G, $<q_0a^n> \Longrightarrow^* <hw>$, where h is the halt state of M. Since we already know that $S \Longrightarrow^* <q_0a^n>$, for every $n \in \mathbb{N}$, we see that in fact $S \Longrightarrow^* <hw>$ for each $w \in L$. We almost have it! We want to show that $S \Longrightarrow^* w$. If we can just get rid of the $<$, the h, and the $>$, we will have that $<hw> \Longrightarrow^* w$ and we can then deduce that $S \Longrightarrow^* w$ for each $w \in L$, as desired. We can do this by adding just a few more rules to G. We want to let the h eliminate the $<$, move through the w, and then eliminate the $>$ along with itself. We need the rules

$$<h \longrightarrow h$$
$$h> \longrightarrow \varepsilon$$

and, for each $\sigma \in \Sigma$,

$$h\sigma \longrightarrow \sigma h$$

We have constructed G so that it generates every string in L. It is not difficult to see that the strings in L are in fact the only strings that are generated by G. That is, L is precisely $L(G)$.

We have now shown, somewhat informally, that a language L is Turing-acceptable if and only if there is a grammar G that generates L. Even though Turing machines and grammars are very different things, they are equivalent in terms of their ability to describe languages. We state this as a theorem:

Theorem 5.2. *A language L is Turing acceptable (equivalently, recursively enumerable) if and only if there is a general grammar that generates L.*

In this section, we have been talking mostly about recursively enumerable languages (also known as the Turing-acceptable languages). What about the Turing-decidable languages? We already know that if a language L is Turing-decidable, then it is Turing-acceptable. The converse is not true (although we won't be able to prove this until the next section). However, suppose that L is a language over the alphabet Σ and that both L and its complement, $\overline{L} = \Sigma^* \smallsetminus L$, are Turing-acceptable. Then L is Turing-decidable.

For suppose that M is a Turing machine that accepts the language L and that M' is a Turing machine that accepts $\overline{L}$. We must show that L is Turing-decidable. That is, we have to build a Turing machine T that decides L. For each $w \in \Sigma^*$, when T is run with input w, it should halt with output 1 if $w \in L$ and with output 0 if $w \notin L$. To do this, T will simulate the computation of both M and M' on input w. (It will simulate one step in the computation of M, then one step in the computation of M', then one step of M, then one step of M', and so on.) If and when the simulated computation of M halts, then T will halt with output 1; since M accepts L, this will happen if and only if $w \in L$. If and when the simulated computation of M' halts, then T will halt with output 0; since M accepts L, this will happen if and only if $w \notin L$. So, for any $w \in \Sigma^*$, T halts with the desired output. This means that T does in fact decide the language L.

It is easy to prove the converse, and the proof is left as an exercise. So we see that a language is Turing-decidable if and only if both it and its complement are Turing-acceptable. Since Turing-acceptability can be defined using other forms of computation besides Turing machines, so can Turing-decidability. For example, a language is Turing-decidable if and only if both it and its complement can be generated by general grammars. We introduced the term "recursively enumerable" as a synonym for Turing-acceptable, to get away from the association with a particular form of computation. Similarly, we define the term "recursive" as a synonym for Turing-decidable. That is, a language L is said to be ***recursive*** if and only if it is Turing-decidable. We then have the theorem:

Theorem 5.3. *Let Σ be an alphabet and let L be a language over Σ. Then L is recursive if and only if both L and its complement, $\Sigma^* \smallsetminus L$, are recursively enumerable.*

Exercises

1. The language $L = \{a^m \mid m > 0\}$ is the range of the function $f(a^n) = a^{n+1}$. Design a Turing machine that computes this function, and find the grammar that generates the language L by imitating the computation of that machine.

2. Complete the proof of Theorem 5.3 by proving the following: If L is a recursive language over an alphabet Σ, then both L and $\Sigma^* \smallsetminus L$ are recursively enumerable.

3. Show that a language L over an alphabet Σ is recursive if and only if there are grammars G and H such that the language generated by G is L and the language generated by H is $\Sigma^* \smallsetminus L$.

4. This section discusses recursive languages and recursively enumerable languages. How could one define recursive subsets of $\mathbb{N}$ and recursively enumerable subsets of $\mathbb{N}$?

5. Give an informal argument to show that a subset $X \subseteq \mathbb{N}$ is recursive if and only if there is a computer program that prints out the elements of X *in increasing order*.

5.3 The Limits of Computation

Recursively enumerable languages are languages that can be defined by computation. We have seen that there are many different models of computation—Turing machines, two-tape Turing machines, grammars, computer programs—but they all lead to the same class of languages. In fact, every computational method for specifying languages that has ever been developed produces only recursively enumerable languages. There is something about these languages—some pattern or property—that makes them "computable," and it is some intrinsic property of the languages themselves, not some peculiarity of any given model of computation.

This is especially interesting since most languages are not recursively enumerable. Given an alphabet Σ, there are uncountably many languages over Σ, but only countably many of them are recursively enumerable. The rest—the vast majority—are not recursively enumerable. What can we say about all these non-recursively-enumerable languages? If the language L is not recursively enumerable, then there is no algorithm for listing the members of L. It might be possible to define L by specifying some property that all its members satisfy, but that property can't be computable. That is, there can be no computer program or Turing machine that tests whether a given string w has the property, since if there were, then we could write a program that lists the members of L.

So, even though almost every language is non-recursively-enumerable, it's difficult to find a particular language that is not recursively enumerable. Nevertheless, in this section we will find one such language. At that same time, we will find an example of a language that is recursively enumerable but not recursive. And we will discover some interesting limitations to the power of computation.

The examples that we will look at in this section involve Turing machines that work with other Turing machines as data. For this to work, we need a symbolic representation of Turing machines—a representation that can be written on the tape of another Turing machine. This will let us create two machines: First, a Turing machine that can generate Turing machines on demand by writing their symbolic representations on its tape. We will design a Turing machine G to do this. And second, a Turing machine that can simulate the computation of other Turing machines whose descriptions are written on its tape.

In order to do all this, we must put some limitations on the states and alphabetic symbols that can be used in the Turing machines that we consider. Clearly, given any Turing machine, we can change the names of the states without changing the behavior of the machine. So, without any loss of generality, we can assume that all states have names chosen from the list: h, q, q', q'', q''', q'''', ... We assume that h is the halt state and q is the start state. Note that there is an infinite number of possible states, but any given Turing machine will only use finitely many states from this list.

As for the alphabets of the Turing machines, I want to look at Turing machines whose alphabets include the symbols 0, 1, a, and of course #. These are the symbols that the machines will use for input and output. The alphabets can also include other symbols. We will assume that these auxiliary symbols are chosen from the list: a', a'', a''', a'''', ... Given a Turing machine whose alphabet includes the symbols 0, 1, a, and #, we can rename any other symbols in its alphabet using names from this list. This renaming will not affect any of the behavior that we are interested in.

Now suppose we have one of these standard Turing machines—one whose states are chosen from the list h, q, q', q'', q''', ..., whose start state is q, and whose symbols are chosen from the list #, 0, 1, a, a', a'', a''', ... Such a machine can be completely encoded as a string of symbols over the alphabet $\{h, q, L, R, \#, 0, 1, a, ', \$\}$. A transition rule such as $\delta(q'', 0) = (a''', L, q)$ can be encoded as a string $q''0a'''Lq$. To encode a complete machine, simply encode each of its transition rules in this way and join them together in a string, separated by \$'s. We now have the symbolic representation for Turing machines that we need.

Note that a string over the alphabet $\{h, q, L, R, \#, 0, 1, a, ', \$\}$ might or might not encode a Turing machine. However, it is a simple matter to check whether such a string is the code for a Turing machine. We can imagine the following process: Generate all the strings over the alphabet $\{h, q, L, R, \#, 0, 1, a, ', \$\}$. Check each string to see whether it encodes a Turing machine. If so, add the string to an output list. In this way, we can generate a list of all strings that encode standard Turing machines. In effect, the standard Turing machines, or at least their symbolic representations, form a recursively enumerable set. Let T_0 be the machine encoded by the first string in this list of standard Turing machines; let T_1 be the machine encoded by the second string; let T_2 be the machine encoded by the third string; and so on. The list T_0, T_1, T_2, ..., includes every standard Turing machine. Furthermore, given $n \in \mathbb{N}$, we can find the symbolic representation for T_n by generating strings in the list until we have $n+1$ strings. Furthermore—and this is the essential point—we can use a Turing machine to do all these calculations. In fact, there is a Turing machine that, when run with input a^n, will halt with the string representation of T_n written on its tape as output. The Turing machine that does this is G, the first of the two machines that we need.

The second machine that we need will be called U. It is a so-called ***Universal Turing Machine***. The single Turing machine U can simulate the computation of any standard Turing machine, T, on any input. Both the symbolic representation of T and that of the input string are written to U's tape, separated by a space. As U simulates the computation of T, it will need some way to keep track of what state T is in and of the position of T on its (simulated) tape. It does this by writing the current state of T on its tape, following T's input string, and by adding a special symbol, such as `@`, to the input string to mark T's position. When U is first started, it begins by adding the `@` to the beginning of the input string and writing a q after the string to represent the start state of T. It is then relatively straightforward for U to simulate the computation of T. For each step in the computation of T, it can determine the current state of T (which is recorded on U's tape) and the symbol which T is currently reading (which is on U's tape, after the `@`). U searches the symbolic representation of T for the rule that tells T what to do in this situation. Using this rule, U can update its representation of T's state, position, and tape to reflect the result of applying the rule. If the new state of T is the halt state, then U also halts. Otherwise, it goes on to simulate the next step in T's computation. Note that when U is given T and an input string w as input, U will halt if and only if T halts on input w. (Obviously, this is a very inefficient simulation, but we are not concerned with efficiency here.)

So, we have our two machines, G and U. After all this setup, we are finally in a position to look at the major theorem that we have been working towards.

Theorem 5.4. *Let T_0, T_1, T_2, ..., be the standard Turing machines, as described above. Let K be the language over the alphabet $\{a\}$ defined by*

$$K = \{a^n \mid T_n \text{ halts when run with input } a^n\}.$$

Then K is a recursively enumerable language, but K is not recursive. The complement

$$\overline{K} = \{a^n \mid T_n \text{ does not halt when run with input } a^n\}.$$

is a language that is not recursively enumerable.

First note that if both K and $\overline{K}$ were recursively enumerable, then K would be recursive, by Theorem 5.3. So, once we show that K is recursively enumerable but not recursive, it follows immediately that $\overline{K}$ cannot be recursively enumerable. That is, the second part of the theorem follows from the first.

To show that K is recursively enumerable, it suffices to find a Turing machine, M, that accepts K. That is, when run with input a^n, for $n \in \mathbb{N}$, M should halt if and only if $a^n \in K$. We can build M from the Turing machines G and U which were introduced above. When started with input a^n, M should proceed as follows: First copy the input. Run G on the first copy of a^n. This will produce a symbolic description of the Turing machine T_n. Now run U to simulate the computation of T_n on input a^n. This simulation will end if and only if T_n halts when run with input a^n, that is, if and only if $a^n \in K$. The Turing machine M that performs the computation we have described accepts the language K. This proves that K is recursively enumerable.

To show that K is not recursive, we need to show that there is *no* Turing machine that decides K. Let H be any Turing machine. We must show that no matter what H does, it does not decide the language K. We must do this without knowing anything more about H that the fact that is it a Turing machine. To say that H decides K would mean that for any $n \in \mathbb{N}$, when H is run with input a^n, H will halt with output 1 if $a^n \in K$ and will halt with output 0 if $a^n \notin K$. To show that H does not decide K we need to show that there is some $n \in \mathbb{N}$ such that when H is run with input a^n, H either fails to halt or else halts but gives the wrong output. Note in particular that we only need to find *one* n for which H does not give the correct result. As we try to find n, we have nothing much to work with but H itself.

To find n, we construct a Turing machine M that is a simple variation on H. When M is run on any input, it duplicates the behavior of H on that input until H halts (if it ever does). At that point, M should check H's output. If H has halted with output 1, then M should go into an infinite loop, so that M never halts in this case. Otherwise, if the output of H is not 1, then M should halt. Now, we can assume that M is one of the standard Turing machines, say $M = T_n$. (If M is not already one of these machines, it is because it uses different names for its states and symbols. Renaming the states and symbols will produce an equivalent machine with the same behavior as M, and we can replace M with this standard machine.)

We now have a Turing machine $T_n = M$ which has the following behavior when it is run with input a^n (note that the n here is the same n as in T_n): If H halts with output 1 on input a^n, then T_n will fail to halt on input a^n. If H halts with output 0 on input a^n, then T_n fails to halt on input a^n. (What T_n might do in other cases is not relevant here.)

Remember that we are trying to show that H does not decide the language K. I claim that, in fact, H does not give the correct answer for

a^n. When H is run with input a^n, it is supposed to halt with output 1 if $a^n \in K$, and it is supposed to halt with output 0 if $a^n \not\in K$. Recall that $a^n \in K$ if and only if T_n halts when run with input a^n.

Suppose that we run H with input a^n. If H does not halt with output 0 or 1, then it has certainly not given the correct answer for a^n. Now, suppose that H halts with output 1 on input a^n. In this case, by the properties of T_n given above, we know that T_n does not halt on input a^n. But that means, by definition of K, that $a^n \not\in K$. By halting with output 1 in this case, H has given the wrong answer for a^n. Finally, suppose that H halts with output 0 on input a^n. We then know that T_n halts on input a^n. But that means that $a^n \in K$. Again, by halting with output 0 in this case, H has given the wrong answer for a^n. So, in no case will H give the correct answer for a^n. This means that H does not decide the language K, because H gives an incorrect answer when it is run with the particular input a^n. H does not decide K, and since H was an arbitrary Turing machine, we see that there is no Turing machine at all that decides the language K. Thus, K is not a recursive language, as the theorem claims.

To decide the language K would be to solve the following problem: Given a Turing machine T_n, decide whether or not T_n will halt when it is run with input a^n. This problem is called the ***Halting Problem***. We have shown that there is no Turing machine that solves this problem. Given the equivalence of Turing machines and computer programs, we can also say that there is no computer program that solves the halting problem. We say that the halting problem is ***computationally unsolvable***.

The halting problem is just one of many problems that cannot be solved by Turing machines or computer programs. In fact, almost any interesting yes/no question that can be asked about Turing machines or programs is in this class: Does this Turing machine halt for all possible inputs in Σ^*? Given this input, will this program ever halt? Do these two programs (or Turing machines) have the same output for each possible input? Will this Turing machine ever halt if it is started on a blank tape? All these problems are computationally unsolvable in the sense that there is no Turing machine or computer program that will answer them correctly in all cases. The existence of such problems is a real limitation on the power of computation.

Index

www.ingramcontent.com/pod-product-compliance
Lightning Source LLC
Chambersburg PA
CBHW071242230926
51849CB00001B/204
* 9 7 8 9 3 9 0 0 1 3 6 2 3 *